International Conference on Civil Engineering and Construction Technology (ICECon 2021)

IOP Conference Series: Materials Science and Engineering
Volume 1229

Online
14 – 15 December 2021

ISBN: 978-1-7138-4687-1
ISSN: 1757-8981

Printed from e-media with permission by:

Curran Associates, Inc.
57 Morehouse Lane
Red Hook, NY 12571

CURRAN ASSOCIATES INC.
proceedings
.com

Some format issues inherent in the e-media version may also appear in this print version.

Institute of Physics
Dirac House, Temple Back
Bristol BS1 6BE UK

Phone: 44 1 17 929 7481
Fax: 44 1 17 920 0979

techtracking@iop.org

Additional copies of this publication are available from:

Curran Associates, Inc.
57 Morehouse Lane
Red Hook, NY 12571 USA
Phone: 845-758-0400
Fax: 845-758-2633
Email: curran@proceedings.com
Web: www.proceedings.com

TABLE OF CONTENTS

Preface

It is a great privilege for us to present the proceedings of ICECon2021, The International Conference on Civil Engineering and Construction Technology 2021 held in Kota Kinabalu, Sabah, Malaysia, from the 14th to 15th of December 2021. The Civil Engineering in Faculty of Engineering, Universiti Malaysia Sabah, organized this Conference; collaborated with the Institution of Engineers (IEM) Sabah, The Association of Consulting Engineers (ACEM) Sabah Chapter, and Green Materials and Advanced Construction Technology Research Unit (GMACT), Universiti Malaysia Sabah.

With the theme "Emerging Technology and Sustainable Construction", ICECon2021 has brought together experts from the industry and academia to disseminate new knowledge, share best practices, expand professional networks, and provide invaluable input for enriching civil engineering and construction technology for the benefit of humankind. Although at times of travel restriction due to the Covid-19 pandemic, over thirty-eight (38) distinguished researchers from three countries (Malaysia, China, and Australia) had virtually gathered and presented papers for this Conference. Two keynote speakers highlighted Renewable Energy and Sustainable Engineering. Their presentations provided a good insight into the latest developments in Green Liquid Biofuels and the Environmental Behaviour of Nanoparticles. Because of the enthusiastic response from the engineering community, the sizeable number of papers prompted the committee to organize parallel sessions during the Conference.

Efforts taken by the peer reviewers contributed to improving the quality of papers provided constructive critics comments; thus, improvements and amendments by authors are greatly appreciated. The success of the ICECon2021 Conference was indeed a collective effort by the international/national advisory committee, session chairs, student volunteers and university financial and administrative staff assistance.

The Civil Engineering Programme and Green Materials and Construction Technology (GMACT), Universiti Malaysia Sabah, has selflessly contributed to hosting the Conference in Kota Kinabalu, Sabah, Malaysia. Without the team's skills, the Conference could not have been a technical success and an enjoyable experience. I would also like to express my most tremendous appreciation to our co-organizers, the Keynote Speakers, the authors, distinguished reviewers, and participants that have supported the ICECon2021 Conference.

Last but not least, we are thankful to the support of the IOP Conference for assisting us in this journey.

ICECon2021 committees
Professor Ir. Dr. Abdul Karim Mirasa, Conference chair
Assoc. Prof. Ir. Dr. Nurmin Bolong, Vice chair
Ts. Dr. Lillian Gungat, Vice chair
Ir. Dr Habib Musa Mohammad, Technical
Assoc. Prof. Dr. Abu Zahrim Yaser, Publication
Dr. Sheikh Mohd Iqbal S. Zainal Abidin, Publication
Ms. Janice Lynn Ayog, Publicity and IT
Ir. Jess Vun Wey Tyng, Publicity and IT
Sr. Ts. Dr. Asmawan Sarman, Local/Event manager
Dr. Hidayati Asrah, Secretariat

Mr. Lim Chung Han @ Andrew, Secretariat
Ms. Adriana Erica Amaludin, Secretariat

IOP Conf. Series: Materials Science and Engineering 1229 (2022) 011002 doi:10.1088/1757-899X/1229/1/011002

Peer Review Statement

All papers published in this volume have been reviewed through processes administered by the Editors. Reviews were conducted by expert referees to the professional and scientific standards expected of a proceedings journal published by IOP Publishing.

- **Type of peer review:** Double Anonymous
- **Conference submission management system:** Morressier
- **Number of submissions received:** 46
- **Number of submissions sent for review:** 39
- **Number of submissions accepted:** 20
- **Acceptance Rate (Submissions Accepted / Submissions Received × 100):** 43.5
- **Average number of reviews per paper:** 2
- **Total number of reviewers involved:** 24
- **Contact person for queries:**
  Name: Nurmin Bolong
  Email: icecon@ums.edu.my
  Affiliation: Universiti Malaysia Sabah

IOP Conf. Series: Materials Science and Engineering 1229 (2022) 012001 doi:10.1088/1757-899X/1229/1/012001

Strength or Ductility Seismic Design? A case study in Sabah using the Eurocode 8 Malaysian National Annex

M.F.T. Ngui[1,*] and D.T.W. Looi[2]

[1] Faculty of Engineering, Universiti Malaysia Sabah, 88400 Kota Kinabalu, Sabah, Malaysia.

[2] Faculty of Engineering, Computing and Science, Swinburne University of Technology Sarawak Campus, 93350 Kuching, Sarawak, Malaysia.

* mftngui@hotmail.com

Abstract. This paper shares the experience of a building project which piloted seismic design using Eurocode 8, with the Malaysian National Annex legislated in 2017. The 23-storey reinforced concrete building features geometric irregularity with podium and two towers, located in Kota Kinabalu, the capital city of Sabah, Malaysia. Sabah had experienced a magnitude-6 earthquake in 2015, where the epicentre Ranau was just 50 km away from Kota Kinabalu. Conventional Probabilistic Seismic Hazard Assessment has indicated that Kota Kinabalu has a peak ground acceleration (PGA) on rock of only 0.04g for a 475-year return period, which appears to be counter-intuitive for such a potentially destructive near-field magnitude 6 earthquake. Hence, four possible scenarios are postulated, i.e., PGA of 0.04g, 0.08g and 0.12g designed with ductility class low (DCL), assuming force-based non-dissipative mechanism and 0.12g designed with ductility class medium (DCM), engaging seismic detailing to form plastic hinges. This paper is written to give local engineering practitioners, who have either no prior experience in seismic design or have been using the basic equivalent force method, to develop a better sense of response spectrum seismic analysis, code-based design and choice of strength or ductility seismic design in low-to-moderate seismicity regions.

1. Introduction

Eurocode 8 (EC8) Part 1 National Annex (NA) has been legislated in Malaysia (MS EN1998-1:2015 [1]) in 2017 [2]. A large part of Malaysia has low seismic activities (minimum 0.07g as recommended in [3]), but not for Sabah. This is evident by a magnitude-6.2 earthquake that occurred at Lahad Datu (East Coast of Sabah) in 1976, and more recently, a magnitude-6 earthquake occurred at Ranau (West Coast of Sabah) in 2015. While it may be appropriate to assume that potentially destructive inter-plate earthquakes do not occur locally in low-to-moderate seismicity regions with intra-plate earthquakes environments, it is well-known that these unpredictable low probability but high consequence events could be equally disastrous, given the unpreparedness of building stocks in these regions. Hence, it is inevitable for the local authority to enforce earthquake design for buildings in Sabah. However, many engineering practitioners in Malaysia have either no prior experience in seismic design or have been using the basic equivalent force method. Looi et al. [4-5] had demonstrated the use of an improved equivalent static method on typical medium-rise reinforced concrete (RC) buildings of 10 stories for code compliant design. Still, the engineering practitioners lack of informed use of higher tier seismic analysis technique, i.e., response spectrum analysis for taller buildings.

Given the challenges, this paper shares the experience of a building project featuring seismic design using EC8 as a case study in Sabah, Malaysia, as an example (see Figure 1). The pilot project is a 23-storey RC building that features geometric irregularity with a podium and two towers, which are structurally designed as equivalent frames. Four possible peak ground acceleration (PGA) scenarios are postulated, i.e., 0.04g, 0.08g and 0.12g designed with ductility class low (DCL), assuming force-based non-dissipative mechanism and 0.12g designed with ductility class medium (DCM), engaging seismic detailing to form plastic hinges. This paper provides insights into the findings, including global drift performance, ultimate global strength demand, ultimate local member strength check, and resource efficiencies (in terms of material tonnage), comparing the four design scenarios.

Figure 1. A case study RC building in Sabah.

2. Seismic design for low-to-moderate seismicity regions

Mainstream earthquake engineering worked examples focus on seismic hazard engineering design principles for higher seismic regions, where ductility design is emphasised [6]. Correspondingly, the default values of the National Determined Parameters (NDP) in the EC8 framework are much characterised by higher seismic practice. Directly adopting the default values intended for high seismic practice may lead to an overly conservative design. Hence, NDP are left to be determined by the nation. Intriguingly, Looi et al. [7] pointed out three challenges faced in determining the NDP during the drafting of the Malaysian NA, which are (i) uncontrolled use of probabilistic seismic hazard assessment (PSHA) technique in producing seismic hazard map, (ii) out-of-date EC8 site classification scheme and (iii) mandating ductile detailing for sites with a higher level of seismic hazard. The authors make use of this pilot project in Sabah to address these challenges.

2.1. Challenge (i)

The reference PGA on rock (a_{gR}) was presented in a zonation map found in the Malaysian NA, which was derived from the conventional PSHA technique for a return period of 475 years. Figure 2 replicated from the Malaysian NA [2] shows a_{gR} is 0.04g for Kota Kinabalu. Sabah had experienced a magnitude-6 earthquake in 2015, where the epicentre Ranau was just 50 km away from Kota Kinabalu. The low a_{gR} value appears to be counter-intuitive for such a potentially destructive near-field magnitude-6 earthquake. The authors also noticed that there are arguments of PSHA maps internationally, more so for low-to-moderate seismicity regions where past earthquake records are scarce [8]. Hence, the varying PGA scenarios studied in this paper are to understand the consequence of resource efficiency by raising the safety level of the building. Nonetheless, it should be reminded that the actual cost is about casualties, repair, and the downtime of the building [9].

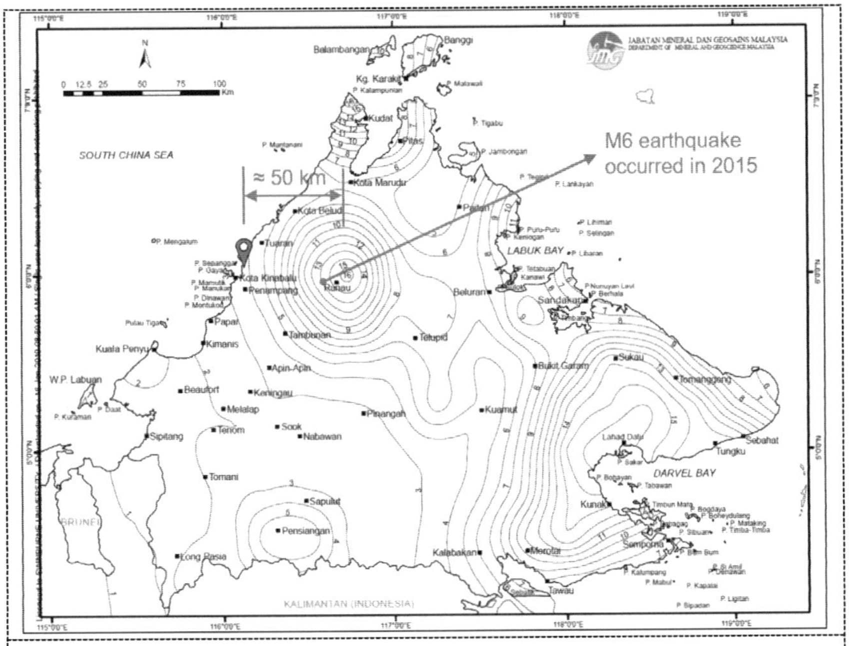

Figure 2. Seismic zonation map of Sabah [2].

2.2. Challenge (ii)

The Malaysian EC8 NA presented with two sets of response spectra plots, i.e., Model A represents the conventional EC8 response spectrum model, which is outdated [10-12], and Model B is the site model with site period parameterisation [13]. The confusion arising from the two sets of response spectra can be rationally resolved with a comprehensive comparison [14].

2.3. Challenge (iii)

It is noted that EC8 mandated ductile detailing requirements for a higher level of seismic hazard as shown in the PSHA contour maps. Consequently, structures located in low seismicity areas are designed with DCL in alignment with the common practice of RC design using Eurocode 2 [15]. On the contrary, for areas that exceed the threshold of low seismicity, designers and builders are forced to adopt ductile design and detailing practices with DCM or Ductility Class High (DCH) consistent with requirements of high seismicity regions. Hence, two of the scenarios postulated in this study, i.e., DCL and DCM for PGA of 0.12g, aim to understand the feasibility of DCL and DCM implementation. The choice of DCH is ruled out as it may be just "over-killing" for low-to-moderate seismicity regions [16-17].

3. Design schemes of the structural systems

3.1. Description of the site

The location of the project site at Kota Kinabalu, the capital of Sabah, Malaysia. Many structural engineers in Malaysia are not familiar with seismic design. Notably, Singapore has published an example for tall buildings [18] based on SS EN1998-1 (2013) [19]. However, it is noted that the response spectrum for Singapore has only considered the long-distance earthquake coming from Sumatra [3]. Unlike Singapore, Kota Kinabalu is subject to a more complex and active seismic environment, compounding the challenge for this pilot project.

The boreholes log data was compiled to determine the seismic ground type. Typically, the depth of the site down to seismic rock level (i.e., standard penetration test, SPT-N > 50) is about 30 m. For response spectra Model A, judgement was based on typical SPT-N of 15 to 50 obtained in most boreholes. For Model B, a common empirical SPT-N to shear wave velocity (V_s) conversion [20] shows that average V_s is about 153 m/s. Readers are recommended to refer to more advanced conversion [21]. Annex A in the Malaysian EC8 NA recommended that the site natural period (T_s) can be estimated using quarter-wave-length theory, which arrived at 0.67 s. Hence, Models A and B results indicated that the site belongs to Ground Type C. Figure 3 shows the plot of normalised response spectra of Ground Type C for Models A and B. It is observed that the demand of Model B is higher at a shorter period (< 1.3 s), and the demand of Model A is higher at a longer period (> 1.3 s).

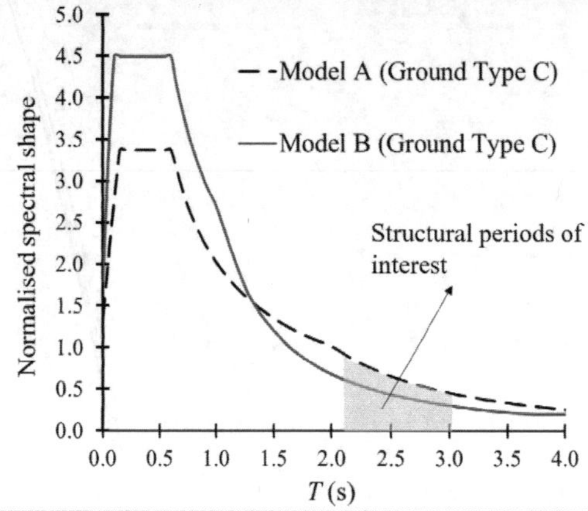

Figure 3. Normalised response spectra for Sabah.

3.2. Description of the case study building

The 23-storey RC building features geometric irregularity with a podium, i.e., 2-level retail space and 5-level car park, and two towers, i.e., a 15-storey tower comprising offices and a 10-storey tower comprising of a 200-room hotel (see Figure 4). The dimension of the building is 67 m in width (X-direction), 76 m in length (Y-direction) and 75 m in height. The gross floor area is about 49308 m^2. The typical story height is 3.4 m, and the thickness of slabs is 150 mm. The dimensions of beams range from 300 × 750 mm to 400 × 900 mm (width × depth), and columns range from 900 mm × 900 mm to 1100 mm × 2550 mm. The core walls have curtailed thicknesses along with the building height, ranging from 230 mm to 250 mm.

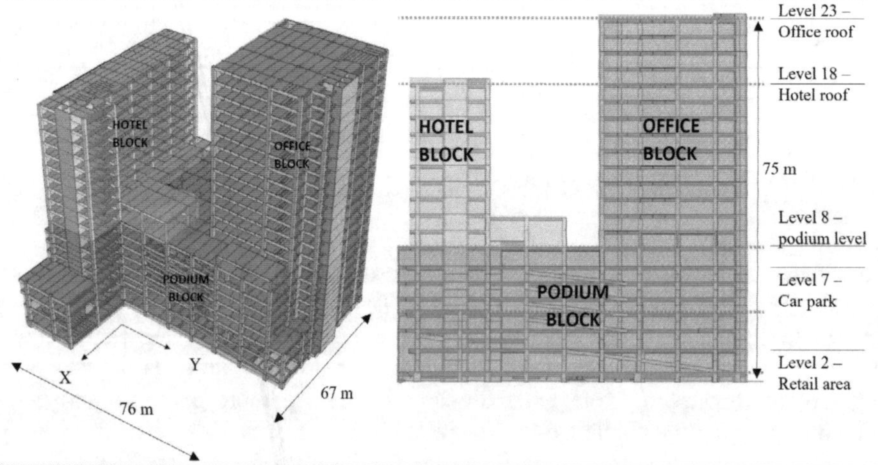

Figure 4. Overall function and dimension of building.

IOP Conf. Series: Materials Science and Engineering 1229 (2022) 012001 doi:10.1088/1757-899X/1229/1/012001

The material used is concrete with a grade of C25/30 and reinforcement with the characteristic yield strength of 500 MPa. Eurocode 1 (MS EN1991-1-1, 2010 [22]) was used to estimate the design load, yielding a volumetric building density of approximately 4.5 kN/m^3. Wind load is calculated according to the local wind code [23].

3.3. Modelling and response spectrum modal analysis

The lateral structural system of the building consists of moment frames and core walls. Figure 5 shows the structural framing, modelled and analysed using a commercial three-dimensional finite element program software [24]. The frames are modelled as line elements, core walls as shell elements and typical floor slabs as shell elements with rigid diaphragms. An average cracked stiffness of 50% was assumed for the concrete members in seismic analysis. The supports are modelled as pinned. The building is classified as Importance Class III in EC8 [1-2].

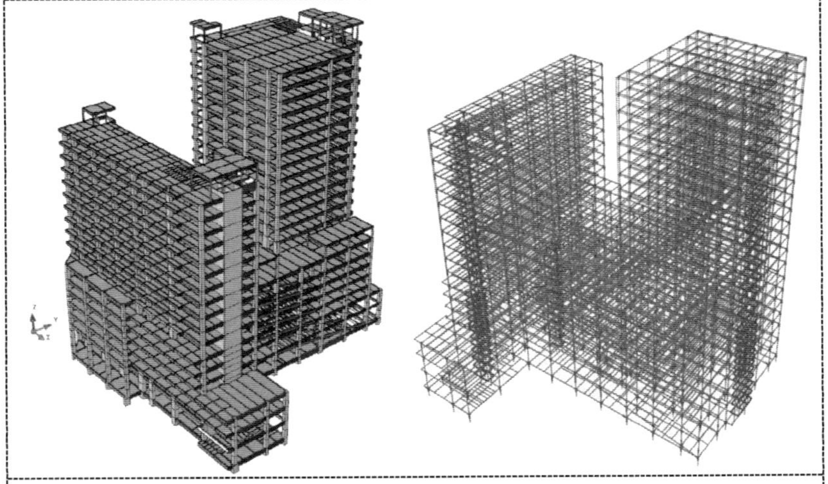

Figure 5. Structural modelling of the building.

Modal analysis was executed to obtain the dynamic properties of the building, i.e., fundamental periods, mode shapes and mass participation factors. The design seismic mass is taken as 1.0 for permanent actions and 0.33 for variable actions. The total number of vibration modes considered is 60 to exceed 90% cumulative mass participation. Figure 6 shows the first three mode shapes of the structure on the plan view, with fundamental periods ranging from 2.2 s to 3.0 s. The first mode has a dominant movement in the Y-direction; the second mode is dominant in the X-direction, and the third mode features torsional behaviour. It is noted that the two towers are far apart; hence there is no inter-tower pounding.

The structural periods of interest (i.e., the first three modes) are also highlighted in Figure 6. In this project, the authors have opted for the conservative response spectra, i.e., Model A. However, readers are strongly recommended to refer to Looi et al. [14] to better understand the site models for their limitations and strengths. The behaviour factor (q) used for horizontal seismic actions are 1.5 for DCL and approximately 3.0 for DCM. The overstrength factor (α_u/α_1) is assumed as 1.3 for DCM, and a reduction factor of 0.8 is applied given the vertical irregularity. CQC modal combination was considered in the results, which caused the loss of negative signs. Therefore, the earthquake direction was considered for reversal loading scenarios in the X and Y-direction, with a 100/30 directional combination rule.

IOP Conf. Series: Materials Science and Engineering 1229 (2022) 012001 doi:10.1088/1757-899X/1229/1/012001

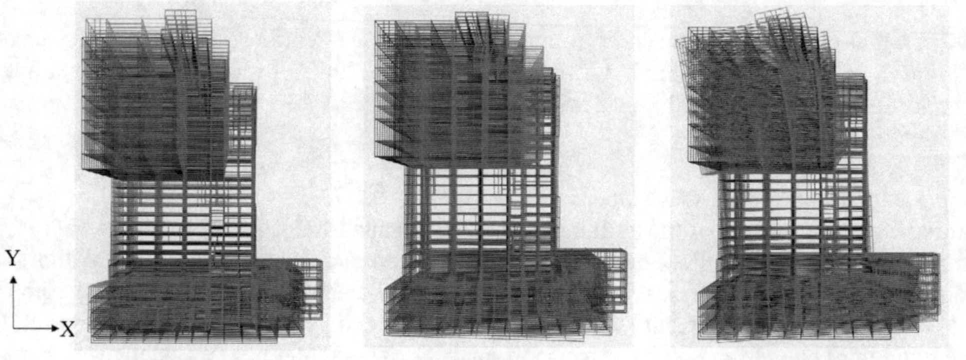

Mode 1 (dominant in X-translational) $T_1 = 3.0$ s

Mode 2 (dominant in Y-translational) $T_1 = 2.7$ s

Mode 3 (dominant in torsional) $T_1 = 2.2$ s

Figure 6. Results of modal analysis shown in plan view.

3.4. Design workflow for scenarios comparison in this study

There are four seismic models considered in this study, i.e., three DCL models subject to PGA of 0.04g, 0.08g and 0.12g, and a DCM model subject to PGA of 0.12g. Additionally, a control model with static loads (permanent, variables, notional and wind actions) designed to RC code (EC2 [15]) has been carried out. A four-stage design workflow is briefly elaborated here. In Stage 1, the building structure is designed to withstand static loads elastically before proceeding to seismic checking. In Stage 2, each model is analysed separately to assess the preliminary member sizes with an iterative process of upsizing and downsizing of members dimensions and minor structural layout re-configurations. Figure 7 shows an example of the analysis contour of the numerical finite element model. In Stage 3, each model undergoes a detailed design process. The outcome is a finished structural model complete with reinforcement (and seismic DCM detail for the model subject to PGA of 0.12g). In the final stage, an independent checker undertook the design review process to validate assumptions, limitations, compliance, long-term performance, fit for purpose, and seismic risk.

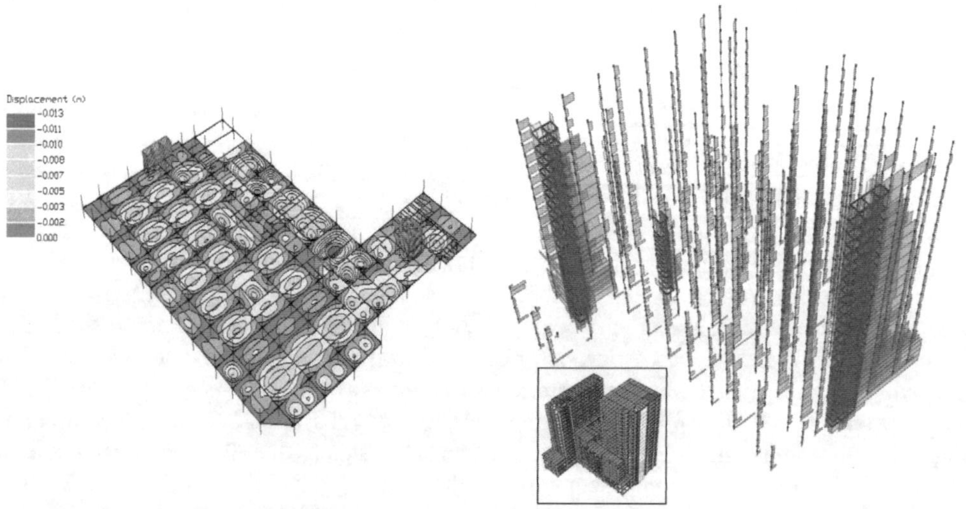

Figure 7. Numerical model analysis contour.

4. Seismic analysis and design results

4.1. Structural analysis results: Drifts

The global structural behaviour (e.g., top roof drift, inter-storey drift, storey shear force and overturning moment) should be first examined before detail elemental design. Figure 8 shows the top roof drift of the building for the control static load case and four seismic cases in X and Y-directions. The maximum recorded deflection is about 300 mm for the DCM case. Superimposed vertical dashed lines are 0.2% wind drift limit (correspond to building height divide by 500) and 0.5% seismic drift limit proposed for low-to-moderate seismicity regions [25]. It is observed that the building is stiff, and all models passed the drift limit requirement. Interestingly, the two models subject to PGA of 0.12g but varied in terms of detailing categories (DCL and DCM) have different drift demands. This subtle but important detail is often overlooked by engineers who are unfamiliar with seismic analysis. Brief explanations are given in this paper for clarity. It is noted that the behaviour factors are different (i.e., q = 1.5 for DCL and ≈ 3.0 for DCM), and the DCM model is essentially subject to a much lower design response spectrum. However, linear elastic analysis is operated in structural software, i.e., lower force yields smaller deformation. Hence, engineers are reminded to check the deflection results for DCM, which must be multiplied by q to arrive at increased deformation demands due to the assumption of enhanced deformation capacity using DCM.

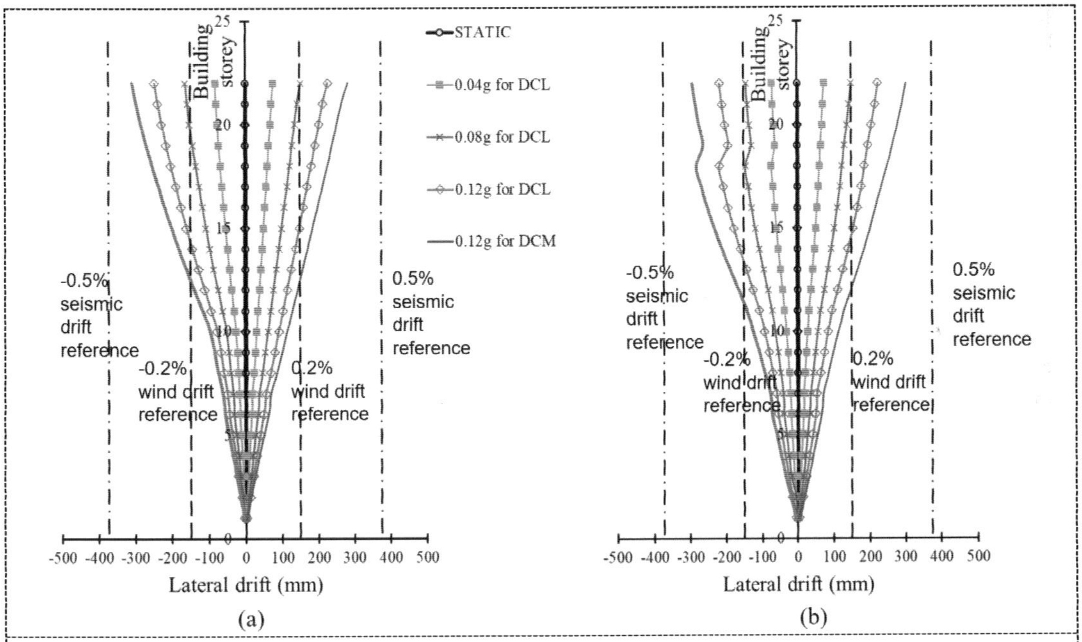

Figure 8. Top roof drifts (a) X-direction (b) Y-direction.

Figure 9 shows the inter-storey drift ratio (IDR) of the models for X and Y-directions. The maximum IDR is about 0.75%. It is observed that none of the models exceeds the reference IDR limit of 1.0% [25] overlaid on the graph. The kink at storey seven is common, largely affected by the interference of podium level [26].

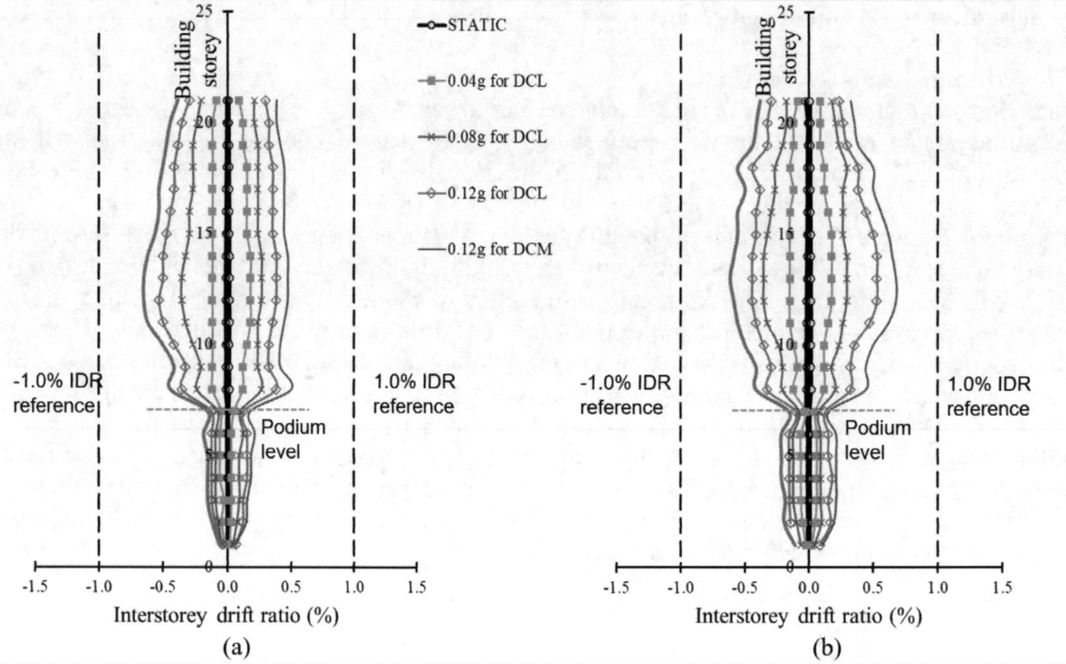

Figure 9. Interstorey drift ratio (a) X-direction (b) Y-direction.

In EC8, the no collapse performance objective is based upon a design return period of 475 years. Given that the code was drafted in the mid-1990s, the concept of Maximum Considered Earthquake (MCE) has not been incorporated into its underlying design philosophy. Structural designers operating in low-to-moderate seismicity regions are cautioned that using the importance factor to increase the ground motion intensity from 475 to 2475 years can exceed 1.5 [27]. Lam et al. [28] and Looi et al. [3] suggested a minimum seismic loading of 0.07g PGA for Malaysia (except Sabah) and 0.12g for Sabah, based on two-third (reciprocal of 1.5) of an MCE of 2475 years. Hence, the authors suggested that the serviceability check be considered for the models with very low PGA, e.g., 0.04g, to avoid cracks in non-structural components.

4.2. Structural analysis results: Global forces

The base shear taken by the frame is between 50% to 65% of the total base shear, which categorised the building as a frame-equivalent dual system as per EC8 [1]. The story shear force (see Figure 10) and overturning moment diagram (see Figure 11) are two important benchmarks in building design. Engineers are cautioned that decisions should not be made by simply adopting the DCM model as it gives smaller global forces. This is because DCM requires much stringent ductile detailing at plastic hinges' locations and capacity design (i.e., strong shear weak bending and strong column weak beam philosophies), which may increase material and workmanship costs.

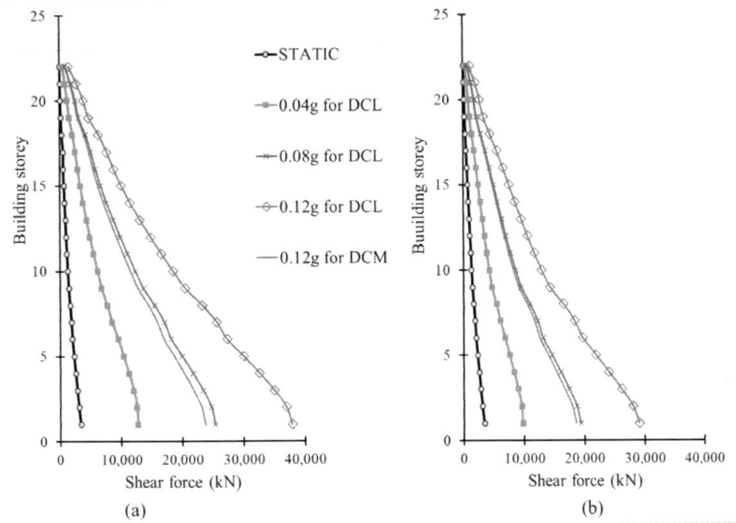

Figure 10. Storey shear (a) X-direction (b) Y-direction.

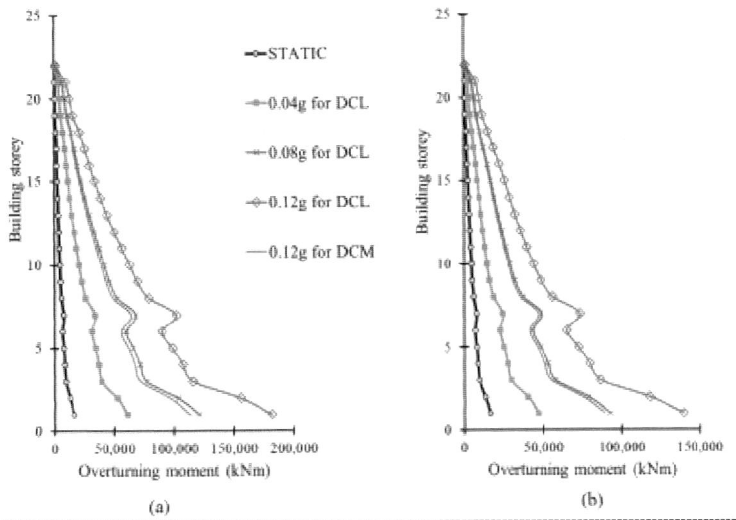

Figure 11. Overturning moment (a) X-direction (b) Y-direction.

4.3. Structural analysis results: Local element forces

Interestingly, given the small global seismic deformation demand and relatively high global seismic force demand as evident in Figures 8 to 9 and Figures 10 to 11, respectively, the use of strength to trade-off ductility can be a feasible structural solution. DCL approach may be sufficient, rather than venturing into DCM. However, two important checks for lateral resisting members (columns and walls) are emphasised regardless of DCL or DCM.

Firstly, brittle failure such as shear crushing should be avoided. Figure 12 demonstrates the maximum shear stress computed for the most critical column and wall of the 0.12g DCL model. The locations of the members are indicated in the building model. The EC2 shear stress limit is superimposed on the graph. The results suggested that the sizing of the column and wall are adequate.

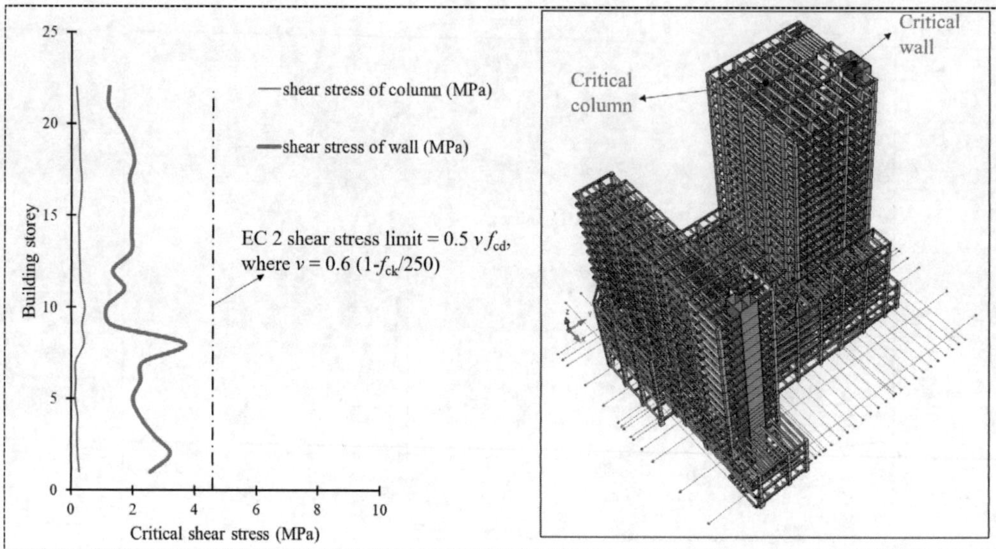

Figure 12. Shear stress for the most critical column and shear wall for the combination of 0.12g DCL.

Secondly, experimental research conducted in the past on non-seismically detailed RC columns [29-30] and non-seismically detailed RC shear walls [31-32] have identified the importance of controlling the axial compression values of these members to give buffer in drift capacity. Although EC2 [15] has not imposed an axial compression limit for column or shear wall, it is worth checking these members using the limits stipulated for DCM in EC8 [1]. Figure 13 depicts the normalised design axial force for the most critical column and shear wall for the 0.12g DCL model. It is observed that the column is not overly compressed, but the wall is highly compressed at storey eight and at the base. As the structure type is considered an equivalent frame, a slight violation of shear wall being overstressed is deemed acceptable.

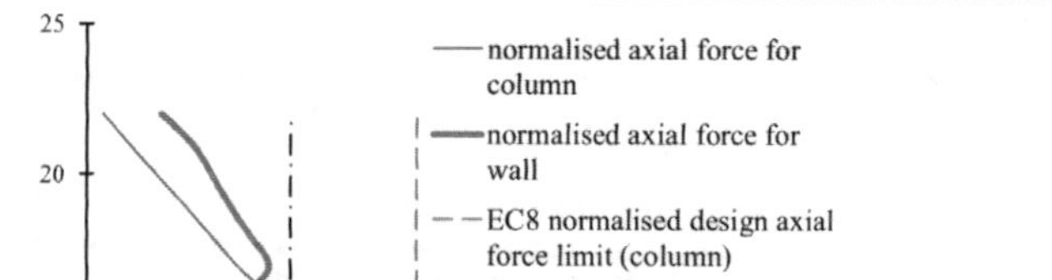

Figure 13. Normalised design axial force for the most critical column and shear wall for the combination of 0.12g DCL.

4.4. Resource efficiency

The material usage for the four seismic models and the control static loading model are compiled and shown in Figure 14. Interestingly, there is not much difference between the control model with the 0.04g DCL model, where only the steel consumption is increased by about 180 tonnes. It is expected to have an initial cost increase in construction material, as seismic design in low-to-moderate seismicity regions is a measure to raise the safety level of buildings [9].

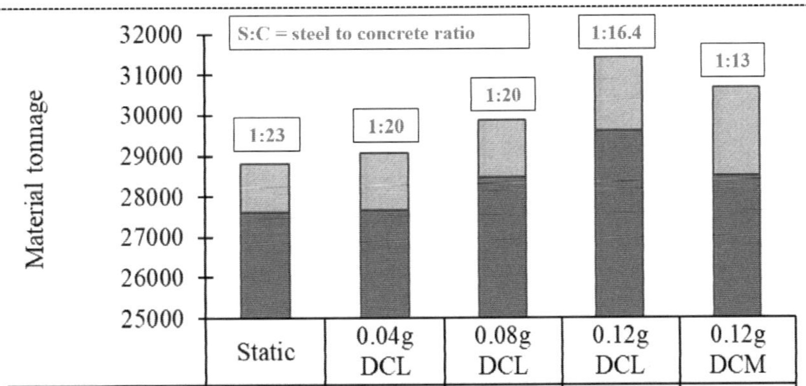

	Static	0.04g DCL	0.08g DCL	0.12g DCL	0.12g DCM
steel (tonne)	1200.7	1379.3	1411.7	1800.9	2192.2
concrete (tonne)	27603.8	27676.5	28466.7	29609.0	28485.1

Types of analysis

Figure 14. Comparison of resource efficiency.

When the seismic level is doubled up to 0.08g and designed to DCL, an increase of 2.9% in concrete consumption (about 800 tonnes) and 2.4% in steel consumption (only about 30 tonnes) is observed, compared to the 0.04g DCL model. The steel to concrete usage ratio for these models is approximately 1 to 20. Hee et al. [33] arrived at a similar conclusion, where the increase is about 2% to 8% for buildings in Sabah. It should be reminded that by increasing 2% to 3% in the initial material costs in this project, the return of investment is a building with two times (200%) safer performance in an earthquake event.

Using strength to trade-off ductility design, the 0.12g DCL model showed an increase in concrete consumption (addition of about 7%) and higher steel consumption (about 30% higher) than the 0.04g model. The steel to concrete ratio is about 1 to 16. On the contrary, using ductility design, the 0.12g DCM model implied a similar concrete consumption, but the steel consumption has substantially increased by 59% compared to the 0.04g DCL model. The steel to concrete ratio is about 1 to 13. It is worth noting that the concrete consumption is slightly higher in the case of 0.12g DCL than 0.12g DCM, which may well be because DCL is essentially an elastic solution governed by geometric stiffness, i.e., a larger member size is needed to resist the seismic actions. Hence, this case study reinforced the argument that DCM may not automatically be the economical solution for low-to-moderate seismicity regions. Engineers are reminded that the increase in initial material cost for DCM has not factored in the time and cost for skilled workers to tie congested rebar cages at plastic hinge locations. Moreover, builders in lower seismic regions are unfamiliar with the stringent detailing works.

In a nutshell, designing to 0.04g using DCL is obligatory as per the seismic hazard map. However, blindly following the stipulated 0.04g is not justified for the magnitude-6 Ranau earthquake, which is about 50 km away from the project site. Designing to 0.08g using DCL appears to be a rational option, as the steel rebar and concrete tonnage are not much different compared to the case of 0.04g. Designing to 0.12g using DCL imposes a penalty on member sizes; on the contrary, DCM imposes a severe penalty on steel consumption.

5. Conclusions

This paper shares the experience of a building project which piloted seismic design using Eurocode 8 Malaysian National Annex. The building is demonstrated for four seismic design scenarios, i.e., PGA of 0.04g, 0.08g and 0.12g designed with DCL, assuming force-based non-dissipative mechanism and 0.12g designed with DCM, engaging seismic detailing to form plastic hinges. A few important points are highlighted, including choice of PGA, insights of the response spectrum, choice of DCL or DCM and some tips to look out for in the input and output of response spectrum analysis. Importantly, engineers are cautioned to check for global behaviour of the building and local shear and axial forces limit of columns and walls. In this case study, it appears that adopting ductility seismic design may not be the best solution, as there will be a penalty on steel utilisation. The authors understand that it is challenging for the industry to elevate the safety level directly to 0.12g overnight, right after the enactment of the Malaysian EC8 NA, where it is the first time the nation has a seismic design code. Hence, taking a mid-way of 0.08g with DCL seems reasonable at this stage, as the rebar tonnage is not much different compared to 0.04g, with only some additional concrete consumption. Nonetheless, the safety level should be progressively upgraded to 0.12g when the stakeholders are ready. It is imperative for the stakeholders to be informed that the actual cost is about casualties, repair, and the downtime of the building after an earthquake event, and not entirely the initial construction cost.

References

[1] MS EN 1998-1-1 2015 *Eurocode 8: Design of structures for earthquake resistance Part 1: General rules, seismic actions, and rules for buildings*. Department of Standards Malaysia
[2] NA to MS EN 1998-1-1:2015 2017 *Malaysian National Annex to Eurocode 8: Design of structures for earthquake resistance Part 1: General rules, seismic actions, and rules for buildings*. Department of Standards Malaysia
[3] Looi DTW, Tsang HH, Hee MC and Lam NTK 2018 *Earthquakes and Structures* **15**(1) 67
[4] Looi DTW, Lam NTK, Tsang HH and Hee MC 2015 Seismic analysis in the low to moderate seismicity region of Malaysia based on the draft handbook. *In Proceedings of the 10th Pacific Conference on Earthquake Engineering*, Sydney, Australia
[5] Looi DTW, Lim EP and Hee MC 2018 Chapter 7: A case study of code compliant design of buildings in regions of low-to-moderate seismicity. Professional Guide: PG-002, *Guideline on Design of Buildings and Structures in Low-to-moderate Seismicity Countries* (in Lam & Chan (eds.)). Chinese National Engineering Research Centre for Steel Construction, The Hong Kong Polytechnic University, Hong Kong, 99
[6] Bisch P, Carvalho E, Degee H, Fajfar P, Fardis M, Franchin P, Kreslin M, Pecker A, Pinto P, Plumier A, Somja H, Tsionis G 2012 Eurocode 8: Seismic Design of Buildings Worked examples. *JRC Scientific and Technical Reports*, European Commission Joint Research Centre, Luxembourg.
[7] Looi DTW, Tsang HH and Lam NTK 2019 The Malaysian Seismic Design Code: Lessons learnt. *In Proceedings of the 2019 Pacific Conference on Earthquake Engineering*, Auckland, New Zealand.
[8] Mulargia F, Stark PB and Geller RJ 2017 *Physics of the Earth and Planetary Interiors* **264** 63
[9] Looi DTW 2019 Earthquake risk in Malaysia: A myth or an underrated truth?, *Campus and Beyond article*, Borneo Post Online, https://www.theborneopost.com/2019/11/20/earthquake-risk-in-malaysia-a-myth-or-an-underrated-truth/
[10] Pitilakis K, Riga E and Anastasiadis A 2012 *Bulletin of Earthquake Engineering* **10** 1377
[11] Pitilakis K, Riga E and Anastasiadis A 2013 *Bulletin of Earthquake Engineering* **11** 925
[12] Pitilakis K, Riga E, Anastasiadis A, Fotopoulou S and Karafagka S 2018 *Soil Dynamics and Earthquake Engineering* **126** https://doi.org/10.1016/j.soildyn.2018.03.03
[13] Tsang HH, Wilson JL, Lam NTK and Su RKL 2017 *Soil Dynamics and Earthquake Engineering* **92** 36

[14] Looi DTW, Tsang HH, Lam NTK and Hee MC 2018 Chapter 12.2: Site classification scheme and response spectrum models for Malaysia. Professional Guide: PG-002, *Guideline on Design of Buildings and Structures in Low-to-moderate Seismicity Countries* (in Lam & Chan (eds.)). Chinese National Engineering Research Centre for Steel Construction, The Hong Kong Polytechnic University, Hong Kong, 255
[15] MS EN 1992-1-1 2010 *Design of concrete structures. General rules and rules for buildings*. Department of Standards Malaysia
[16] Booth E 2014 Future Directions for Eurocode 8. *Second European Conference on Earthquake Engineering and Seismology*, Istanbul, Turkey Aug. 25th-29th
[17] Bisch P 2018 Eurocode 8: Evolution or Revolution? *In Proceedings of the Recent Advances in Earthquake Engineering in Europe* (in Pitilakis K. (ed.)), 16th European Conference on Earthquake Engineering – Thessaloniki
[18] BC3 2013 *Guidebook for Design of Buildings in Singapore to Design Requirements in SS EN-1998-1*. Singapore: Building and Construction Authority
[19] SS EN1998-1 2013 *Eurocode 8: Design of structures for earthquake resistance – Part 1: General Rules*, Seismic Actions and Rules for Buildings. SPRING, Singapore
[20] Imai T and Tonouchi K 1982 Correlation of N value with S-wave velocity and shear modulus. *In Proceedings of the 2nd European Symp. on Penetration Testing*, Amsterdam, 67
[21] Hu YW, Lam NTK, Khatiwada P, Menegon JS and Looi DTW 2021 *CivilEng* **2**(3) 712 https://doi.org/10.3390/civileng2030039
[22] MS EN 1991-1-1 2010 *Actions on structures Part 1-1: General actions - densities, self-weight, imposed loads for buildings*. Department of Standards Malaysia
[23] MS 1553 2002 *Code of Practice on Wind Loading for Building Structure*. Department of Standards Malaysia
[24] ProtaStructure 2018 Prota Software User's Manuals. Prota Asia Sdn. Bhd. Kuala Lumpur, Malaysia
[25] Su RKL, Looi DTW, Tang TO and Law CW 2014 Performance Based Seismic Design for Tall Buildings in Hong Kong. *In Proceedings of the Presentation at HKIE/IStructE Joint Structural Division Annual Seminar 2014 Advances in Seismic Engineering*, Jockey Club Auditorium, The Hong Kong Polytechnic University, Hong Kong
[26] Yacoubian M, Lam NTK, Lumantarna E and Wilson JL 2017 *Engineering Structures* **148** 639
[27] Tsang HH 2014 Seismic performance requirements and collapse risk of structures, *In Proceedings of the Presentation at HKIE/IStructE Joint Structural Division Annual Seminar 2014* Advances in Seismic Engineering, Jockey Club Auditorium, The Hong Kong Polytechnic University, Hong Kong
[28] Lam NTK, Tsang HH, Lumantarna E and Wilson JL 2016 *Earthquakes and Structures* **11**(4) 539
[29] Wilson JL, Wibowo A, Lam NTK and Gad EF 2015 *Australian Journal of Structural Engineering* **16**(1) 62
[30] Raza S, Tsang HH and Wilson JL 2018 *Magazine of Concrete Research* **70**(21) 1081
[31] Su RKL and Wong SM 2007 *Engineering Structures* **29**(8) 1957
[32] Looi DTW, Su RKL, Cheng B and Tsang HH 2017 *Engineering Structures* **115** 312
[33] Hee MC, Lam NTK, Tsang HH, Looi DTW, Zaeem AH and Lim EP 2016 Summary Update of Cost Implication on Proposed Malaysian NA for EC8 on Office Buildings and Link Houses, *JURUTERA* (the monthly bulletin of The Institution of Engineers, Malaysia). January Issue 42

Acknowledgments
The authors acknowledged Pet Royston Paulus Tinus for his contribution as the lead seismic modeller. The consent by Pacific Sanctuary Holdings Sdn. Bhd. in sharing the project as a case study in this paper is gratefully acknowledged. The authors thank Eramaju Synergy Sdn Bhd for the case study and

design resources and Arkitek Billings Leong & Tan Sdn. Bhd. for their consent to use the architect impression images.

Surface Waviness Evaluation of Two Different Types of Material of a Multi-Purpose Hall Using Terrestrial Laser Scanner (TLS)

I Neza[1], M I Mohamed[2] and W M Syafuan[1,3]

[1] Faculty of Engineering, Universiti Pertahanan Nasional Malaysia, Kuala Lumpur 54000, Malaysia.

[2] Faculty of Engineering, Universiti Malaysia Sabah, Sabah 88400, Malaysia.

[3]Corresponding author: wmsyafuan@upnm.edu.my

Abstract. The construction industry is rapidly integrating current technology and new construction techniques to increase productivity and efficiency in construction operations. Quality control and progress tracking are labour-intensive and time-consuming operations. Several surface waviness assessment methods have been created to alleviate the drawbacks of one assessment method over the other, focusing on measuring the quality of concrete slabs. Current methods of assessing waviness using one-dimensional (1D) survey lines produce inaccurate results and are prone to mistake. It does not reflect the actual condition of the entire floor concerning waviness. Two objectives of this study are to 1) evaluate the feasibility of using Terrestrial Laser Scanner for surface waviness assessment using slope analysis of different types of material and 2) propose a valuate pattern and image for different slope surface analyses. Leica RTC 360 laser scanner was used to scan the floor of the Multi-purpose Hall. The plan image of the floor slab and one of the 3D point clouds were recorded. Slope analysis shows a consistent pattern of slope distribution for floors with tiles. This is because the tile used concrete screed during installation to have a smooth surface during construction. The distribution of slope analysis of tile surface mostly between 0 to 0.0625% while the slope analysis slope variation of laminated floor between 0% - 0.5%. The distribution shows that floor variation with the laminated wood surface is high, especially between 0 to 0.5%. From the result obtained, this method has the benefit of combining rich 3D data from TLS for fast and reliable surface waviness assessment during construction. This technology has massive potential in the construction sector because of its capacity to record as-built conditions consistently, fast, and accurately.

1. Introduction

Many industries are being transformed by emerging technologies focusing on improving product quality, increasing production rates, and lowering long-term costs. The construction industry is rapidly integrating current technology and new construction techniques to boost productivity and efficiency in construction operations. The construction industry's digitisation is being accelerated by incorporating these tools and technologies into construction operations. Project field engineers can track the progress of construction work using accurate as-built data received from ongoing construction projects.

Quality control and progress tracking are labour-intensive and time-consuming operations. Failure to communicate project performance with all involved parties effectively and timely leads to delays and cost overruns in construction work [1]. Furthermore, data produced manually for progress tracking may not be readily available for analysing project performance. It may be inferred that traditional tracking techniques in projects are unreliable and prone to error. Quickly identifying any differences enables necessary actions to reduce the impact of a delay on the building workflow.

Dimensions Quality Control (QC) ensures that items are manufactured to the prescribed dimensional tolerances. As-built dimensions of cast-in-place concrete elements often differ from those specified initially in as-designed plans [2]. It is critical to conduct prompt inspections after the completion of concrete elements, especially in terms of flatness and waviness. The failure to detect the imperfections in newly constructed surfaces during the earlier stages of construction causes delays in carrying out necessary repair works [3]. The repair, demolition, removal, and replacement of defective concrete elements entail additional costs that could amount to as much as 12% of the project contract value [4-7].

Several surface waviness assessment methods have been created to alleviate the drawbacks of its practical use, focusing on measuring the quality of concrete slabs. The F-number approach, for example, was created to address the drawbacks of the Straightedge method, while the Waviness Index (WI) method was created to address the drawbacks of the F-number method. The lack of data linked with each of these methods makes it difficult to have a clearer picture of how elevations and undulations vary over time. Current methods of assessing waviness using one-dimensional (1D) survey lines may not accurately reflect the actual state or waviness of the entire floor. Bosché and Guenet [8] proposed integrating BIM and TLS to capture as-built elements that comply with the specified surface flatness tolerances using 3D point clouds obtained using TLS and data from BIM. Two objectives of this study are to:

- Evaluate the feasibility of using TLS for surface waviness assessment using slope analysis of different types of material, and
- Proposed a valuate pattern and image for different slope surface analyses.

2. Dimensional Quality Control in the Construction Industry

When it comes to concrete slabs, various elements influence the dimensional quality of cast-in-place concrete slabs, including hot temperatures, location, and finishing procedures used during construction. Waviness in concrete slabs causes problems that are not only unsightly but also hinder the efficiency of machinery such as lift trucks or very narrow aisle (VNA) vehicles. It also affects the health and safety requirement of sports facilities such as futsal and basketball court. The waviness of concrete slabs in industrial facilities, such as large warehouses, must be thoroughly examined because failure to detect waviness and deviations from specified tolerances can significantly impact the operational activities that the floor is designed to handle [9]. The effect of abnormalities in the floor on the stability of VNA vehicles is depicted in figure 1. Static lean (s) of VNA trucks are caused by differences in elevation between the left and right wheels (d). Because of the waviness of concrete slabs, the static lean can rise by ten times [10].

2.1. Current Methods of measuring concrete slab waviness

Surface imperfections are divided into two categories: "abrupt" and "gradual." Offsets and fins generated by displaced or misaligned formwork, loose knots, and other faults in formwork materials are examples of abrupt abnormalities that must be checked by direct measurement. Gradual imperfections must be verified using a straight template for level surfaces or a suitable counterpart for curved surfaces, measuring 3.0m for unformed surfaces and 1.5m for formed surfaces. A 3mm deviation from a 1.5m straight edge placed in any position on a presumably planar surface is commonly specified as a flatness tolerance. The existing methods for detecting concrete slab waviness in the construction sector require extensive human interaction, are labour intense and time-consuming. The Straightedge method puts a 3.0-meter Straightedge across a survey line on the floor and uses a stainless-steel slip gauge to measure the distance between the Straightedge and the floor [11]. The

IOP Conf. Series: Materials Science and Engineering 1229 (2022) 012002 doi:10.1088/1757-899X/1229/1/012002

procedure of laying out the Straightedge over vast surface areas is time-consuming and results in random measurement mistakes [12]. This time-consuming method only offers information on the variances between as-built and as-designed locations at a few measured places.

The F-numbers approach was introduced to eliminate random mistakes in measurements by using devices that allow for the measurement of elevation changes at fixed intervals, resulting in more accurate findings. The F-number approach produces flatness numbers (Floor Flatness (FF) and Floor Levelness (FL) values) that are difficult to interpret. It gives two numbers to work that is FF and FL. The flatness of the measured floor surface point is described by FF, whereas FL describes the levelness of the measured floor surface point. The measurements are taken at 1-ft intervals along each survey line, as stated in the ASTM E1155-14 standard [13]. The values obtained from many survey lines are statistically processed to provide FF and FL figures that represent the conditions of the whole floor surface.

The WI method identifies different periods of floor undulation between 1.0 and 3.0 meters, which correlate to periods of surface undulation that affect forklift operation [15][16]. The WI method produces results expressed in inches and is, therefore, easier to interpret. Despite having a significant advantage over the Straightedge and F-number approaches, the WI method suffers from limitations. Because these procedures entail manually moving measurement equipment around the floor's surface, the findings are prone to human error. Inaccuracies in measurements are exacerbated by random errors resulting from carelessness when handling measurement tools. The primary methodologies' significant limitation is the difficulty of recovering consistent results across multiple measurement sessions.

$$\frac{d}{D} = \frac{s}{H} \rightarrow s = \frac{d \cdot H}{D}$$

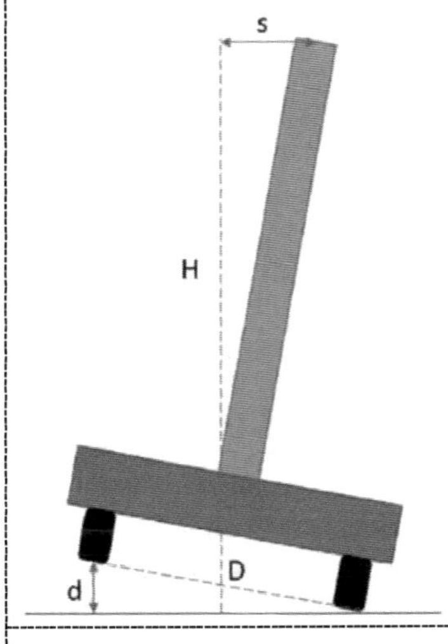

s - Static lean of a VNA truck in mm for D=1.2 m		d - Difference in elevation between the left and right wheels of a fork lift truck [mm]									
		3	4	5	6	7	8	9	10	11	12
H - height of racking [m]	6	15	20	25	30	35	40	45	50	55	60
	7	18	23	29	35	41	47	53	58	64	70
	8	20	27	33	40	47	53	60	67	73	80
	9	23	30	38	45	53	60	68	75	83	90
	10	25	33	42	50	58	67	75	83	92	100
	11	28	37	46	55	64	73	83	92	101	110
	12	30	40	50	60	70	80	90	100	110	120
	13	33	43	54	65	76	87	98	108	119	130
	14	35	47	58	70	82	93	105	117	128	140
	15	37	50	63	75	87	100	113	125	137	150
	16	40	53	67	80	93	107	120	133	147	160
	17	42	57	71	85	99	113	128	142	156	170
	18	45	60	75	90	105	120	135	150	165	180

Figure 1. The effect of the difference in elevation between the wheels on the static lean for a VNA [10]

2.2 Emerging technology for measuring concrete slab waviness

The adaptability of Terrestrial Laser Scanning (TLS) technology has been put to the test in a variety of construction industries applications. TLS is a modern surveying technology gaining traction in the construction industry. TLS has been used, for example, to analyse the conditions of remote sites where human access is difficult or risky [17], to create BIMs that represent the as-built conditions of building

IOP Conf. Series: Materials Science and Engineering 1229 (2022) 012002 doi:10.1088/1757-899X/1229/1/012002

facilities [18], and to monitor construction progress. Accurate project as-built information is required for dimensional quality control measurements to make timely choices.

TLS is gaining popularity for dimensional QC because of its ability to quickly give as-built data in the form of dense and accurate 3D point clouds (with accuracy ranging from sub-mm to mm) [19]. TLS overcomes accuracy and repeatability issues and allows for capturing data representing the geometry of finished surfaces, overcoming the data sparsity restrictions of traditional surveying methods. TLS provides an effective technique for gathering dense as-built data spanning the entire slab surface compared to existing measurement equipment used in current standard flatness measurement methods. Fuchs et al. [20] and Shafer and Weber [21] developed deformation monitoring methods to discover the differences in positions of TLS data points concerning a reference surface when processing TLS data for QC. In [22], a colour map generated from TLS data was used to analyse the flatness of multi-story building facades. Tang et al. [23] devised three techniques for determining the difference in elevation of points in a point cloud concerning a plane derived from a Building Information Modelling (BIM) model or a user-specified plane. The method described in [24] employed an elevation map with different colours representing different height intervals. This method displayed each point's height concerning a reference plane derived from a BIM model. These methodologies and current approaches identified areas with varying degrees of deviation, but they failed to quantify the waviness or duration of surface undulations.

3. Method

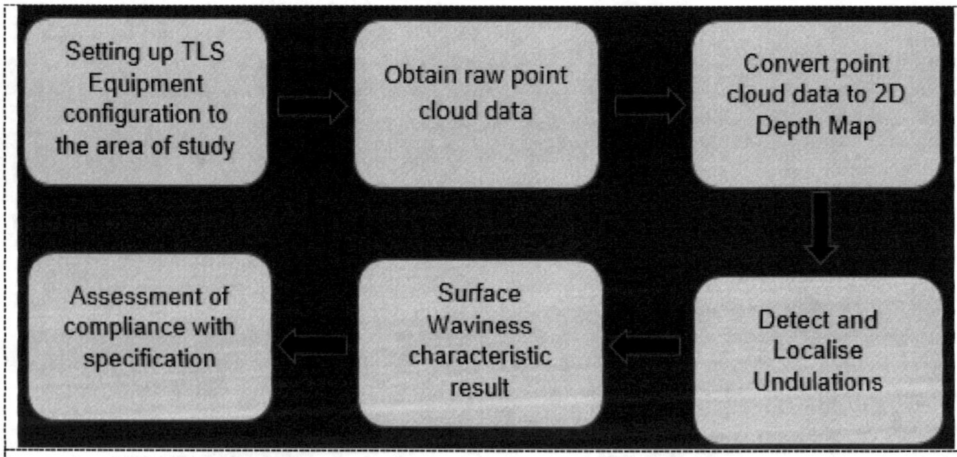

Figure 2. Overview of the research methodology.

As summarised in figure 2, the proposed methodology was used to test the applicability of the proposed framework to dimensions and surface QC of surface flatness. The input is a raw point cloud that may result from the scanning process's co-registration of numerous scans. The registered point cloud data noise is removed using a commercial point cloud processing software's capability, leaving a clean point cloud of the area of interest (i.e. concrete slab). The slab surface's pre-processed point cloud is aligned (parallel) to the XY plane, which is most likely already the case. As a result, z coordinates represent each point's elevation, making subsequent analysis more manageable. The raw 3D point cloud comprises data points from a floor area and its immediate surroundings and structures. Several laser scans are often co-registered using a conventional target-based approach to create the point cloud. The raw point cloud is first pre-processed to isolate the area of interest, the concrete slab, from the rest of the data. The next stage is to create a depth map representing the undulating zones corresponding to several undulations types. A Leica RTC 360 laser scanner was used to scan the warehouse building's floor slab. After putting up the scanner's tripod, an in-situ floor slab from

IOP Conf. Series: Materials Science and Engineering 1229 (2022) 012002 doi:10.1088/1757-899X/1229/1/012002

UPNM's multi-purpose hall was scanned. At 50 metres, the scanner's 3D position accuracy is 3 mm, and at 100 metres, it is 10 cm.

4. Result and discussion

A Leica RTC 360 laser scanner was used to scan the hall floor. The plan image of the floor slab and one of the 3D point clouds recorded are shown in figure 3. In both the plan view and the point cloud, the region of interest with a surface area of roughly 100 m^2 is marked. For the point cloud registration process to be more manageable, six targets were set at various points around the facility. The entire scanning process took around 45 minutes, including setup, scanning, dismantling, and relocating. The raw point clouds, which included noise from laser scans, were initially loaded into commercial point cloud processing software. Approximately 100,000,000 points in the point cloud corresponding to the slab region of interest. Figure 4 shows the data analysis process using Cyclone Register 360 and Cyclone 3DR.

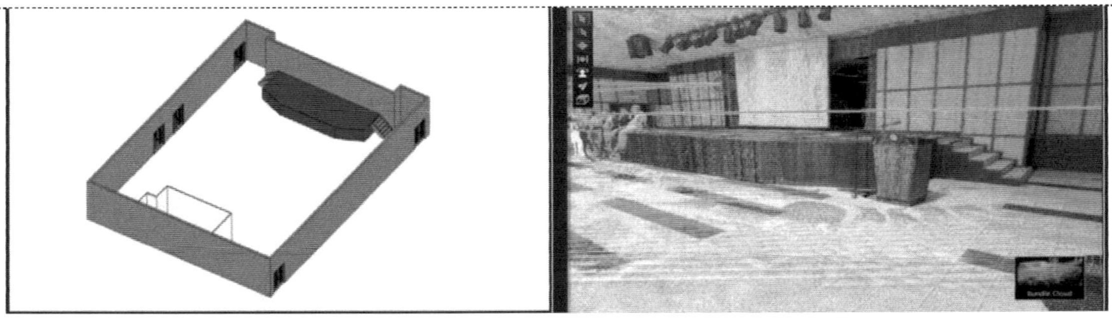

Figure 3. The plan view and the 3D point cloud view of the study area. The stage consists of the laminated wood floor, while the rest uses a tile surface.

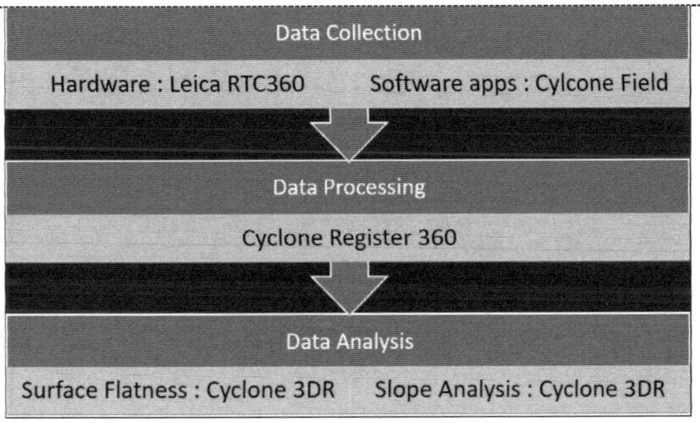

Figure 4. Hardware and software used for data processing and data analysis

The resulting slope analysis is shown in figures 5 and 6. Both surfaces were analysed for 1.0% and 0.5% slope to see the surface flatness using the Cyclone 3DR software. Figure 5(a) shows result for floor with tile surface with slope variation between 0% to 1% while figure 5(b) shows area slope variation between 0% - 0.5%. Both slope analysis shows a consistent pattern of slope distribution. The traditional tile installation methods relied on secondary mortar beds placed over underlying concrete slabs to provide a level and smooth profile for ceramic tile installations. The slope percentage for tile surface is mostly between 0 to 0.0625%, which is well in the specification for concrete flatness (not more than 3mm for every 1500mm).

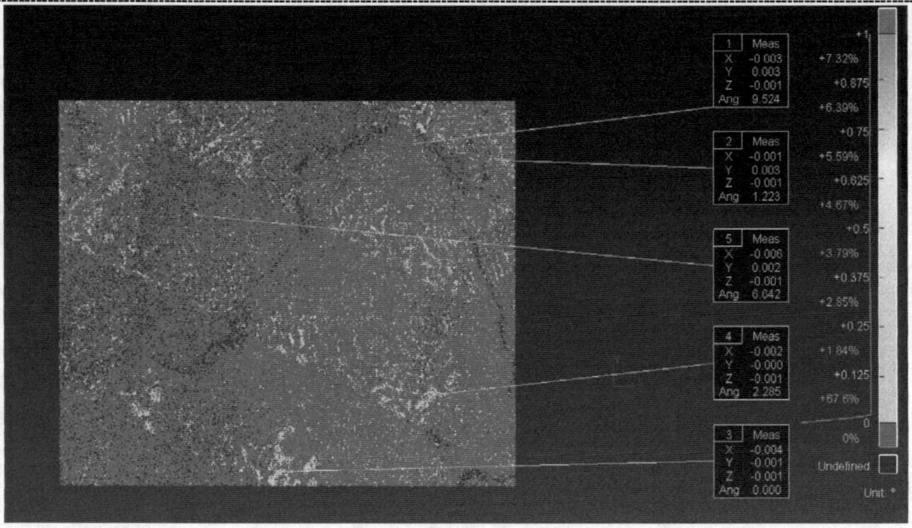

(a)

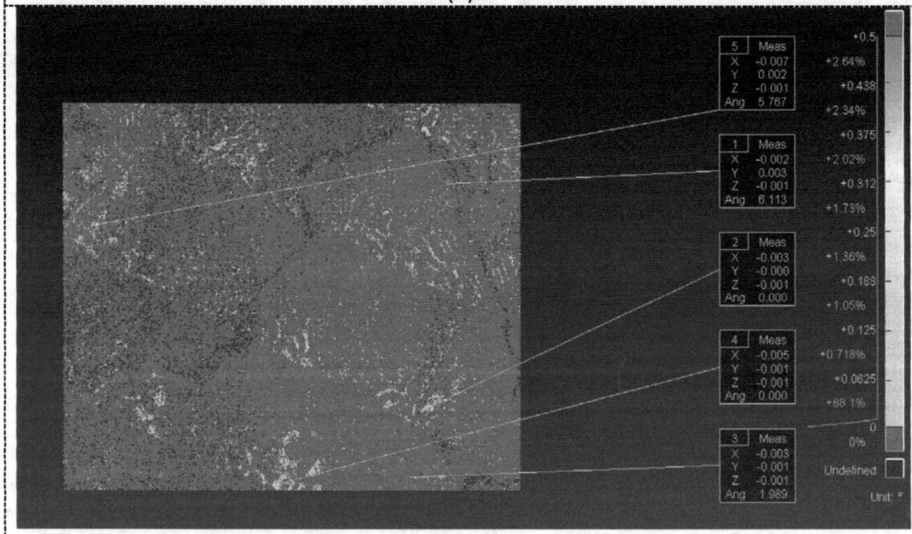

(b)

Figure 5. Slope analysis for floor with tile (a) 1% slope and (b) 0.5% slope.

Figure 6(a) shows the result for the floor with the laminated wood surface with slope variation between 0% to 1%, while figure 6(b) shows area slope variation of laminated wood surface between 0% - 0.5%. The distribution shows that variation of the wood surface is high, especially between 0 to 0.5%. The installation of the laminated wood does not have any material to cater to the waviness of the concrete surface. From the result obtained this method benefits combining rich 3D data from TLS. According to the testing results, the suggested approach provides efficient colour-coded deviation maps to depict the flatness information of construction surfaces or component surfaces. When comparing the two experimental results, it becomes clear that the construction surface of a wood-laminated floor requires further investigation. The proposed method demonstrates an efficient method for quickly locating areas with large deviations or places that deviate from the as-designed model. Because concrete surface waviness quality control is critical for construction, there is a rising need to present an automated waviness evaluation method to replace the time-consuming existing human testing methods.

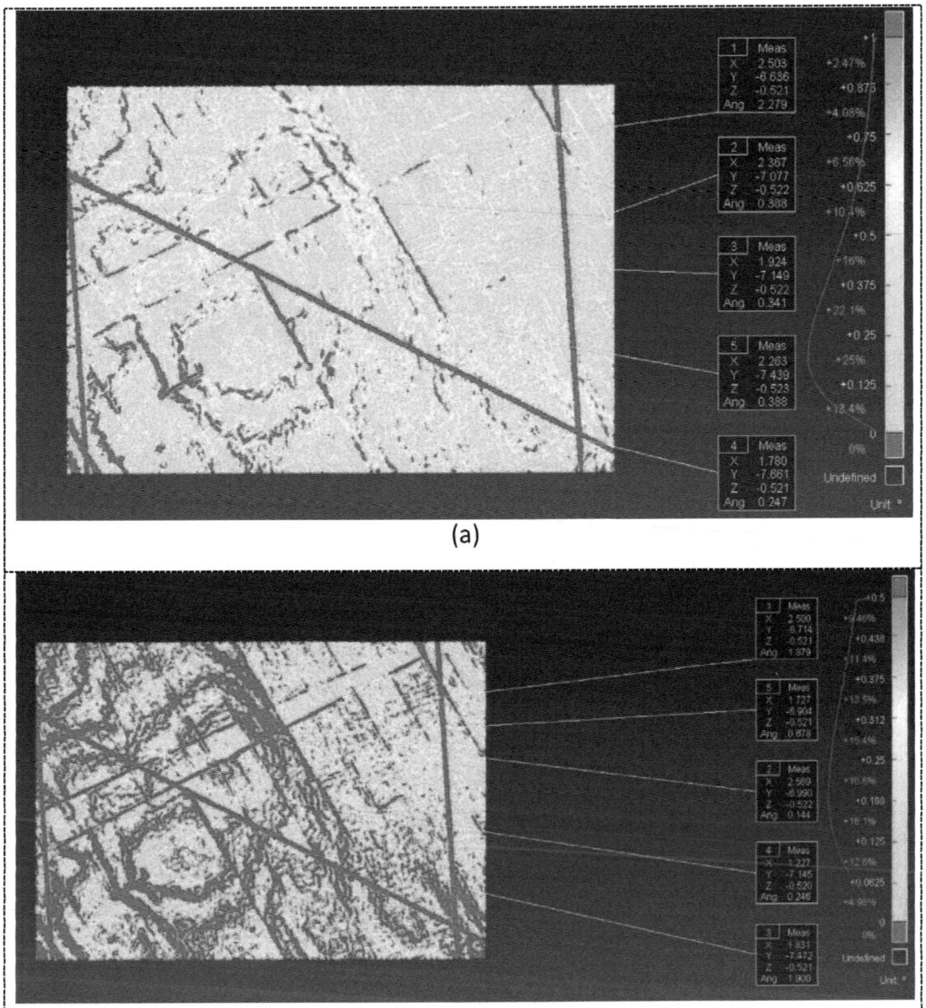

(a)

(b)

Figure 6. Slope analysis for floor with laminated wood (a) 1% slope and (b) 0.5% slope.

5. Conclusions

Because of its capabilities to capture and record as-built elements quickly and accurately, TLS has considerable potential in the construction sector. One such area is dimensional quality control. The dimensional QC analysis provides versatility when looking at surface undulations with a wide range of distinct periods. The distribution of deviations on each surface can be easily analysed, quickly locating the area at the actual location on the surface. In particular, the area with high deviations often appears near the adjacent edges of two surfaces localisation in compliance evaluation and corrective work

IOP Conf. Series: Materials Science and Engineering 1229 (2022) 012002 doi:10.1088/1757-899X/1229/1/012002

planning by highlighting regions on the surface. Future research efforts should focus on enhancing the practicality of using laser scanning to measure the waviness of surfaces. This paper proposed a method for automatically evaluating concrete surfaces and precast concrete component surfaces to illustrate the distribution of waviness information. The technology suggested in this paper can visualise undulations according to various characteristic periods on site.

References

[1] J.D. Owolabi, L.M. Amusan, C.O. Oloke, O. Olusanya, P.F. Tunji-Olayeni, O. Dele, N.J. Peter, I.O. Omuh, "Causes and effect of delay on project construction delivery time," International Journal of Education and Research, vol. 2, no. 4, pp. 197–208, 2014.A reference

[2] D. K. Ballast, Handbook of Construction Tolerances, Second. John Wiley & Sons, 2007.

[3] H. P. Gillette, C. S. Hill, C. T. Murray, H. B. Kirkland, and S. C. Hadden, Engineering and Contracting, no. v. 39. Gillette Publishing Company, 2013.

[4] McKinsey Global Institute, "Reinventing Construction: A Route to Higher Productivity," 2017.

[5] U. S. Census Bureau., "Value of construction put in place at a glance," 2013.

[6] J. L. Burati and J. J. Farrington, "Costs of quality deviations in design and construction," Bureau of Engineering Research, the University of Texas at Austin, 1987.

[7] Patterson and W. B. Ledbetter, "The cost of quality: A management tool," in Excellence in the Constructed Project, 1989, pp. 100–105.

[8] F. Bosché and E. Guenet, "Automating surface flatness control using terrestrial laser scanning and building information models," Automation in Construction, vol. 44, pp. 212–226, 2014.

[9] F. Bosché and B. Biotteau, "Terrestrial laser scanning and continuous wavelet transform for controlling surface flatness in construction – A first investigation," Advanced Engineering Informatics, vol. 29, pp. 591–601, 2015

[10] F. R. Neal, Concrete Industrial Ground Floors. Thomas Telford, 2002.

[11] D. K. Ballast, Handbook of Construction Tolerances, Second. John Wiley & Sons, 2007.

[12] B. M. Phares, G. a. Washer, D. D. Rolander, B. a. Graybeal, and M. Moore, "Routine Highway Bridge Inspection Condition Documentation Accuracy and Reliability," Journal of Bridge Engineering, vol. 9, no. 4, pp. 403–413, 2004.

[13] American Society for Testing and Materials (ASTM), "Standard Test Method for Determining F F Floor Flatness and F L Floor Levelness Numbers (Metric) 1," ASTM, vol. 96, no. Reapproved 2008, pp. 7–14, 2015.

[14] American Society for Testing and Materials (ASTM), "Standard Test Method for Determining Floor Tolerances Using Waviness, Wheel Path," ASTM vol. 04, no. November 1998, pp. 1–12, 2004.

[15] C. N. Ytterberg, "Using the Waviness Index To Improve Floor Flatness," 1996.

[16] R. E. Loov, "Is the F-number system valid for your floor?" Concrete International, 123 vol. 12, no. 1, pp. 68–76, 1990.

[17] G. S. Cheok, W. C. Stone, R. R. Lipman, C. Witzgall, and J. Uisod, "Lidars for construction assessment and update," vol. 9, no. 5, pp. 463–477, 2000.

[18] H. Son and C. Kim, "Advanced Engineering Informatics Semantic as-built 3D modelling of structural elements of buildings based on local concavity and convexity," Advanced Engineering Informatics, vol. 34, no. August, pp. 114–124, 2017.

[19] M.C. Amann, T. Bosch, M. Lescure, R. Myllyla, and M. Rioux, "Laser ranging: a critical review of usual techniques for distance measurement," Optical Engineering, vol. 40, no. 1, pp. 10–19, 2001.

[20] P. A. Fuchs, G. A. Washer, S. B. Chase, and M. Moore, "Applications of LaserBased Instrumentation for Highway Bridges," Journal of Bridge Engineering, vol. 9, no. 6, pp. 541–549, 2004.

[21] T. Schafer and T. Weber, "Deformation measurement using terrestrial laser scanning at the

IOP Conf. Series: Materials Science and Engineering 1229 (2022) 012002 doi:10.1088/1757-899X/1229/1/012002

hydropower station of Gabcikovo," in INGEO 2004 and FIG Regional Central and Eastern European Conference on Engineering Surveying, FIG, Copenhagen, Denmark., 2004.

[22] M. C. Israel and R. G. Pileggi, "Use of 3D laser scanning for flatness and volumetric analysis of mortar in facades," vol. 9, no. 1, pp. 91–106, 2016.

[23] P. Tang, B. Akinci, and D. Huber, "Characterisation of three algorithms for detecting surface flatness defects from dense point clouds," IS&T/SPIE Electronic Imaging, vol. 7239, p. 72390N–72390N, 2009.

[24] N.J. Shih and P.-H. Wang, "Using point cloud to inspect the construction quality of wall finish," Proceedings of the 22nd eCAADe Conference, no. October, pp. 573–578, 2004.128

Multivariate Statistical Analysis (MVA) in Complimenting Building Condition Assessment as a Diagnosis Tool for Reinforced Concrete Structures Rated as Category 3, 4 and 5

J. Pirah[1,2*], J. Lozitin[2], N. Samshuddin[3], R. Roslee[4] and M.F. Zulkipli[1,5]

[1] Oreads Building Care, Sabah, Malaysia.
[2] Alamega Konsult, Sabah, Malaysia.
[3] Konsultan Azam Sempurna, Sabah, Malaysia.
[4] Universiti Malaysia Sabah, Natural Disaster Research Centre (NDRC), Sabah, Malaysia
[5] Jurutera Inovasi, Sabah, Malaysia.

*Corresponding author: jefferyjim@gmail.com

Abstract. In most structural repair works or building condition assessment, structural appraisal has been done through visual inspection for tell-tale signs which highlight the possible defect. This method is extended to the standard guideline by Jabatan Kerja Raya (JKR) in their BCA format in determining the physical condition rating. Consulting engineers or structural specialists are to participate when it is determined that the structure requires material testing in order to come up with diagnoses and prognosis. The use of descriptive statistics in both reconnaissance or desktop studies as well as in standard specifications have limitation in determining the actual structural issues. This study demonstrates the advantages of using Multivariate Statistical Analysis (MVA) in summating the right diagnosis and prognosis for structural repair works.

Keywords: Building Condition Assessment, Multivariate Statistical Analysis, Structural Integrity, Building Pathology, Concrete Testing

1. Introduction

Sabah is a state in Malaysia which is situated at the North of Borneo Island and public schools are plagued with dilapidation issues. Under the 11th Malaysia Plan, the federal government allocates RM1.252 billion for upgrade and redevelop dilapidated schools. This renders the urgency and the magnitude of such unsafe state where 414 schools are considered as unsafe by the State Education Department and District Education Office where 52 schools are considered highly unsafe by the Public Work Department (JKR) [1]. An estimated of 54.5% of schools in Sabah are considered as dilapidated and only 37 schools were repaired and another 83 schools are undergoing repair works [2].

Building condition assessments were conducted by JKR based on Standard Guideline for Building Condition Assessment (BCA) based on visual inspection and the overall physical condition rating of building is based on Building Condition Assessment Online System (BCA-OS) developed by Universiti Teknologi Malaysia. Schools which listed under category 3, 4 and 5 will undergo subsequent material testing as part of the diagnosis stage to determine the robustness of the structural members [3] with the appointment of the consulting engineer [4]. After material tests were conducted and defects were mapped, a diagnosis will be explained in parsimony which usually leads to localized or regionalized repair works. Most often, diagnoses were guided by descriptive statistical analyses which may not be sufficient and usually may lead to least holistic approaches when providing accurate prognosis. In some cases, the outcomes can be futile. In school building condition assessment cases cited by Syahirah *et al.* [3], prognoses were specified, however, a caveat was mentioned where continuous

IOP Conf. Series: Materials Science and Engineering 1229 (2022) 012003 doi:10.1088/1757-899X/1229/1/012003

monitoring should be conducted where tell-tale signs are not sufficient to explain the cracks generated and buildings are still in distress. This requires additional studies to be conducted in order to bridge arise issues and holistic prognosis.

A school in Kunak continues to dilapidate and degrade although it was repaired previously with proper repair works and was supervised by the government's agency. A detailed desktop studies and investigations were conducted in order to provide proper diagnosis and prognosis for this main building structure in order to avoid another structure repair work throughout its designed lifespan which is not cost efficient. Instead of using descriptive statistics as prescribed by many codes of practices, standard specifications and other guidelines, a multivariate statistical analysis (MVA) was used for a full diagnosis. This study will provide insight and bridge the missing link between JKR Standard Guideline for Building Condition Survey, relevant normative references from specifications and codes of practices to a holistic prognosis which is valid and reliable.

2. Literature Review

2.1. Building Pathology

Building pathology is a concept which was not fully embraced when the performance of building is concern as it only forms part of the overall assessment based on multidisciplinary approach. Nevertheless, in order for a conclusive and inclusive in resolving the building as the main structural host for all other installation and facilities, a holistic prognosis is required. This concept requires detailed understanding and knowledge mainly engineering in regards to how buildings are designed, constructed, utilized and the change of form and function, as well as various mechanism which materials and environments which are affected the building itself [5]. With a good understanding of building pathology, a professional engineer can identify, investigate and conduct proper diagnosis before come up with a holistic prognosis of defects and make proper recommendations. This field also includes the study of failures in interrelationship of structures and materials as they coexist in certain environment, function and forms (Hutton & Rostron, 1989 as cited in Watt, 2008 [5]). Most of the time, the basis of assessment is on components and element such as architectural finishing which often discuss about non-structural cracks, dampness and movements. Little emphasis was made for materials unless it is deemed to be necessary as stipulated in JKR Standard Guideline for Building Condition Assessment.

2.2. Standard Specifications

The standard specification and/or code of practice in conducting structural investigation is loosely interpreted with morbid collections of references since the scope of concrete specifications are relatively vast. Nevertheless, it all depends on the visual inspection and desktop studies to establish the investigation working brief in determining the emphasis of material investigation. Since concrete and reinforcement bars are the main materials for most of reinforced concrete schools and buildings, the focus would be the type of test to be deployed in making proper diagnosis. The morbid collection of reference will range from American, British and European standard specifications, detailed testing and references from the American Society for Testing and Materials (ASTM) or International Union of Laboratories and Experts in Construction Materials, System and Structures (RILEM).

For the use of concrete testing in regards to compressive strength, the use of BS EN 13791:2007 – Assessment of in-situ compressive strength in structures and precast concrete components and its local annex, MS1242:1991 – Guide to assessment of concrete strength in existing structure. The other specification series that shall be used in assessing other concrete properties shall be based on BS 1881 – Test for concrete.

2.3. Multivariate Statistical Analysis (MVA)

Typically, most of the engineering practices, including what is prescribed in BS EN 13791 and MS 1242 involves descriptive statistics. Multivariate statistical analysis involves another branch of statistical approach in resolving issue by making proper interpretation through data inference by the mean of conducting data analysis using inferential statistics. The difference between both types of statistics is the purpose of data analysis. Descriptive statistical analysis will give insights on the available dataset of specific samples from a population with numerical analysis and data visualization. This statistical analysis is also considered as univariate. Inferential statistical analysis provides the advantage in making inference, hypothesis testing, determining relationships and making prediction [6]. These features from inferential statistics are powerful tools in making proper diagnosis and a resolving tool in determining the right prognosis. A meta-analysis was conducted and no journal is found to have similarity on research parameters relevant to concrete properties in this particular study.

This particular study involves relationships of outcomes from various testing and the process of summating findings as a prognosis which involves more than just a parameter as well as in just one location.

3. Research Methodologies

When assessment was conducted by JKR, it was found that this repaired school buildings have structural issues despite previous repair and a consulting engineer was called in to examine the actual condition at the site. Material testing are required to be carried out in order to provide summary for repair method which complies with BS EN 1504 – Products and Systems for the Repair and the Protection of Concrete Structures.

The guideline in determining the structural condition of the building is based on BS EN 13791:2007 – Assessment of in-situ compressive strength in structures and precast concrete components and its local annex, MS1242:1991 – Guide to assessment of concrete strength in existing structure. The method of extracting samples for hardened concrete and testing will comply with BS EN 12504-1 – Testing Concrete in Structure – Cored Specimens – Taking, examining and testing in compression.

Table 1. List of Material Testing

Test Detail	Test Descriptions	Test Guidelines
Carbonation Depth	In order to check the chemical condition in the concrete, phenolphthalein will be sprayed at the freshly cored sample. The depth of the carbonation is indicated by the distance and depth covered by the pink colour after applying phenolphthalein	BS EN 14630
Chloride Content	Laboratory test would identify the level of chloride content and/or ingress into the concrete as well as a measure to check if chloride ions are passive inhibitor in the concrete structure.	BS 1881 Part 124
Electromagnetic Covermeter	This test method is deployed in order to check the thickness of concrete structure cover by determining the distance between the surface and the reinforcement bars. Aside from that, this test will ensure coring work to acquire samples will not affect the structure's reinforcement bars.	BS 1881 Part 204
Coring / Sampling and Compressive Strength Test	Concrete samples are extracted from the structural members and the sample frequency shall be guided by BS EN 13791.	BS EN 12504 Part 1
Rebound Hammer	The main method for acquiring the strength of the concrete at the site is by using Rebound Hammer method at selected areas of the building and to be correlated to the cored samples.	BS EN 12504 Part 2.

Half-Cell Potential	A non-destructive test which shall be performed to measure the surface potential which is associated with corrosion current.	BS1881 Part 201 or ASTM C876-80
Concrete Density Test	Test should be performed to all core samples in accordance to determine the density of each concrete sample	BS EN 12390 Part 7
Concrete Hacking	A destructive test which shall be performed in order to evaluate the condition of the concrete and the reinforced steel bars in the said location. Aside from that, this will also provide the actual reinforcement bars' size and number as well as spacing to be tallied to any indicative as-built drawing to form the steel (bending) schedule.	
Petrography Examination	A laboratory test which shall be performed by utilizing concrete cores extracted from selected areas where samples are scrutinized under petrographic microscope.	BS 1881 Part 211 or ASTM C856-04

The other concrete related test in regards to other properties shall be governed by BS 1881 – Testing Concrete with its sub-series of guidelines for different types of concrete properties. The other test conducted also comply with British and European Standards as well as ASTM. The list of destructive and non-destructive tests to be carried out at the site are shown in Table 1.

Once all material test was conducted, multivariate statistical analysis shall be conducted by using IBM SPSS software where dataset acquired from field record shall be tabulated accordingly, cleaned dataset as well as outliers shall be identified and removed as required by MS1242.

4. Result and Discussion

The result from test samples is tabulated and all data shall be cleaned from outliers. The outliers eliminated are as shown in Figure 1 as both readings are out of range based on MS 1242. Once outliers are removed, data can be interpreted as shown in Figure 2.

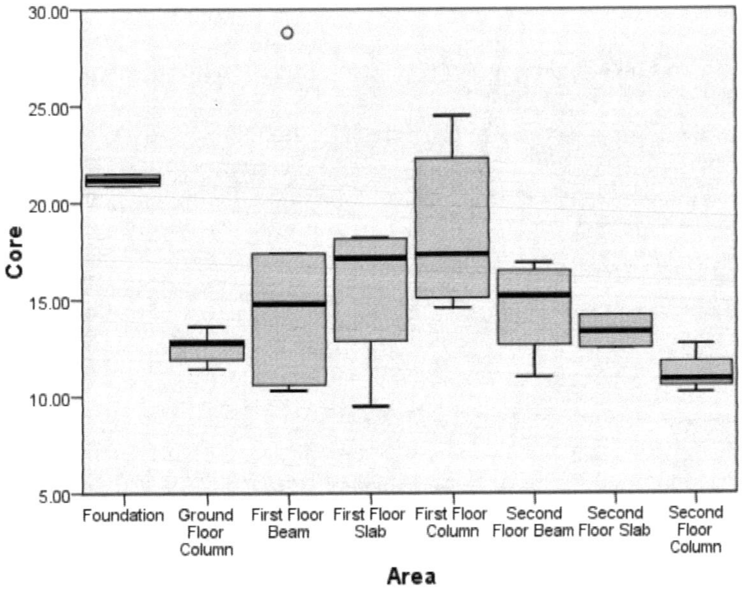

Figure 1. Boxplot for Concrete Core Strength (in kN/m^2) Distribution.

After 2 samples are considered as outliers, the remaining 31 samples shall be the basis in conducting statistical analyses. A descriptive analysis was conducted to project out the data quality received from field material tests and the result are as tabulated in Table 2. The result yield from descriptive renders the mean for concrete characteristic strength of 15.10 kN/m^2 is below the desired characteristic strength of at least 25 kN/m^2. Based on BS EN 13791, the actual characteristic strength after determining the standard deviation and mean for the core result and eliminating all suspected outliers, followings are the final value based on Approach A as per Section 7.3.2 of BS EN 13791 where the estimated in-situ characteristic strength, $f_{ck,is}$ should be the lowest of both Equation (1) and (2).

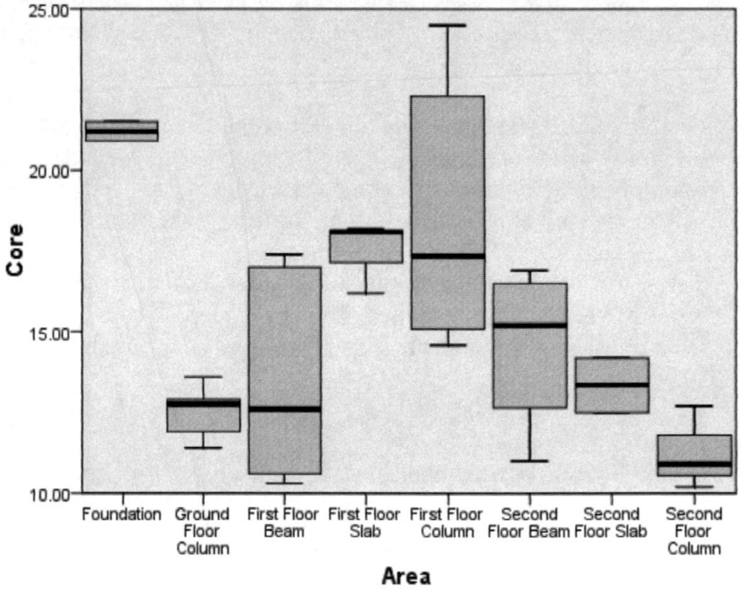

Figure 2. Boxplot for Concrete Core Strength (in kN/m^2)

Based on Equation 1, the mean in-situ characteristic strength $f_{m(n),is}$ is 15.10 kN/m^2, and lowest in-situ characteristic strength $f_{is,lowest}$ is 10.20 kN/m^2; k_2 is 1.48 (6%); S is 3.77 kN/m^2 and hence the result is 9.52 kN/m^2. For Equation 2, $f_{ck,is}$ is 14.20 kN/m^2. Therefore, the in-situ characteristic strength of the concrete is 9.52 kN/m^2 based on Equation 1 and 2.

$$f_{ck,is} = f_{m(n),is} - k_2 \times s \quad (1)$$

$$f_{ck,is} = f_{is,lowest} + 4 \quad (2)$$

$$f_{ck,is} = f_{m(n),is} - t_{0.05} \times s \quad (3)$$

Nevertheless, with reference made to MS 1242 Section 6.2, the method in determining the in-situ strength shall be modified where K_2 or $t_{0.05}$ shall be from the statistical table indicates the value of 1.64 or 1.70. Therefore, the final in-situ strength shall be as calculated based on Equation 3, where characteristic strength, $f_{ck,is}$ value is 8.691 kN/m^2.

Table 2. Descriptive Statistics for Core Sample and Rebound Hammer.

Descriptive Statistics

	N	Range	Min	Max	Mean	Std. Dev	Variance	Skewness		Kurtosis	
	Statistic	Statistic	Statistic	Statistic	Statistic	Statistic	Statistic	Statistic	Std. Error	Statistic	Std. Error
Core	31	14.30	10.20	24.50	15.1000	3.77262	14.233	.774	.421	-.038	.821
Hammer	31	36.50	16.50	53.00	35.0323	8.87312	78.732	.025	.421	-.659	.821

4.1. Bivariate and Univariate Statistical Analysis Using Other Concrete Property Test

After the characteristic strength of the concrete is determined and is below the requirement, other concrete properties are investigated and other tests were conducted. These will allow the consulting engineer to probe deeper the root cause of the problem that may lead to the degradation of concrete strength and reinforcement bars. Without inferential checks involving making cross-tabs and multivariate statistical analyses, it is generally impossible to establish the relationship of strength degradation to time, spatial, function, loadings, stresses and environment which are dimensions that can be measured.

4.1.1. Covermeter.

The covermeter survey results yielded that the mean cover for all samples is 45.70 mm with concrete cover thickness ranging from 11 mm to 85 mm at the site. The bar and link spacing are in acceptable arrangement and spacing. This shall form the spacing for structural members' steel bar schedule.

Table 3. Descriptive Statistics for Covermeter.

	N	Range	Min	Max	Mean	Std. Dev	Variance	Skewness		Kurtosis	
	Stat	Stat	Stat	Stat	Stat	Stat	Stat	Stat	Std. Error	Stat	Std. Error
Concrete Cover	10	74.00	11.00	85.00	45.70	27.45	753.567	.144	.687	-1.593	1.33

4.1.2. Concrete Hacking.

The subsequent destructive test conducted on similar areas revealed mixed results where more than half of reinforcement bars in selected areas are corroded and the remaining are not corroded as shown in Table 5. Nevertheless, most of the corroded reinforcement bars have a diameter of 90% or more than the initial bars which are designed for. This conclude that some of the reinforcements may not need to be replaced unless during the execution of structural repair works. Nevertheless, some of these steel reinforcements lost more than 90% of the initial diameter size accounted for the design and it is found to be impractical to splice, then these reinforcement bars may need to be replaced as a whole.

Table 4. Descriptive Statistics for Hacking and Corrosion Check.

	N	Range	Minimum	Maximum	Mean	Std. Dev	Variance	Skewness		Kurtosis	
	Statistic	Statistic	Statistic	Statistic	Statistic	Statistic	Statistic	Statistic	Std. Error	Statistic	Std. Error
Hacking	10	1.00	.00	1.00	.6000	.51640	.267	-.484	.687	-2.277	1.334

IOP Conf. Series: Materials Science and Engineering 1229 (2022) 012003 doi:10.1088/1757-899X/1229/1/012003

Table 5. Cross-Tabulation – Hacking * Concrete Cover Thickness.

		Concrete Cover							Total
		11.00	25.00	30.00	40.00	65.00	75.00	85.00	
Hacking	No Corrosion	1	0	0	1	0	1	1	4
	Corrosion	1	1	1	1	1	1	0	6
Total		2	1	1	2	1	2	1	10

Figure 3 and Figure 4 render the outcomes of cover thickness in affecting reinforcement bars' condition and defectiveness of structures. Figure 3 summarizes that corrosion occurred in a similar distribution as per non-corroded reinforcement bars. However, the boxplot indicates that non-corroded reinforcement bars have higher average cover thickness than those which are corroded. The other finding as shown in Figure 4 indicates that the defectiveness of the concrete is not necessarily depends on cover thickness.

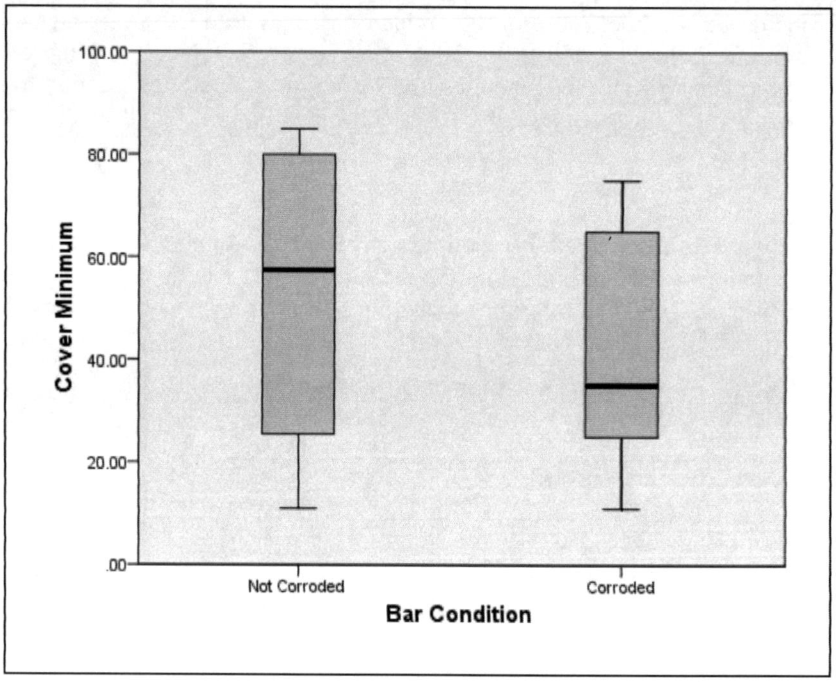

Figure 3. Boxplot – Minimum Cover (in mm) vs. Bar Condition

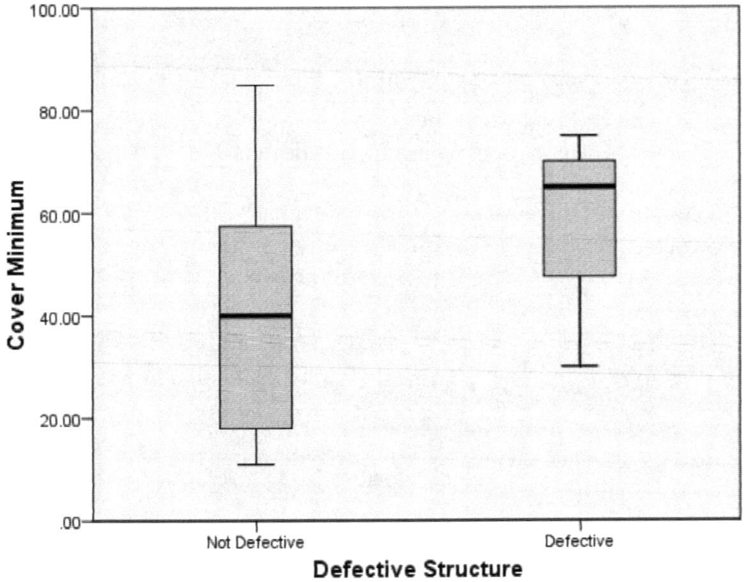

Figure 4. Boxplot – Minimum Cover (in mm) vs. Defective Structure

4.1.3. Half-Cell Potential.

Another confirmatory test has been conducted in order to validate the actual condition of steel reinforcements at areas (not similar areas as of covermeter survey) have 6 sample areas with 5% corrosion possibility and the remaining 4 sample areas with 50% corrosion possibility and 95% corrosion possibility. Most of the corroded areas are exposed to weather while covered areas have no corrosion to top slab or beam except for the soffit. This construed to the finding at the site that most of the concrete components close to the link-way and the link-way itself has the potential of corrosion due to splashing during precipitation. Table 6 until Table 10 summarize the state of corrosion within sampled structural members. Structural members with the highest possibilities of corrosion are columns on the first floor and the second floor. Although reinforcement bars are corroded, one of the two samples from column structure did not show any sign of being defective.

Table 6. Cross-Tabulation Half-Cell Components * Location.

		Half-cell Location				**Total**
		Ground Floor	First Floor	Second Floor	Roof	
Half-Cell Component	Column	1	2	1	0	4
	Beam	0	2	1	1	4
	Slab	0	0	2	0	2
Total		1	4	4	1	10

Table 7. Cross-Tabulation Half-Cell Result * Half-Cell Components.

		Half-Cell Components			Total
		Column	Beam	Slab	
Half-Cell	5% Possibility	2	2	2	6
	50% Possibility	0	2	0	2

	95% Possibility	2	0	0	2
Total		4	4	2	10

Table 8. Cross-Tabulation Half-Cell Result * Half-Cell Locations.

		Half-cell Location				Total
		Ground Floor	First Floor	Second Floor	Roof	
Half-Cell	5% Possibility	1	1	3	1	6
	50% Possibility	0	2	0	0	2
	95% Possibility	0	1	1	0	2
Total		1	4	4	1	10

Table 9. Cross-Tabulation Half-Cell Result * Defective Structure.

		Defective Structure		Total
		Not Defective	Defective	
	5%	4	2	6
Half-Cell Potential	50%	0	2	2
	95%	1	1	2
Total		5	5	10

Half-cell potential result provides the insight of the possibility to debunk the common myth where exposure to environment or ambient is key in causing degradation to concrete structure. Half-cell reading is usually low when the structure is in the state of being passive; however, it does not clearly relate defectiveness of a concrete structure to the corrosion of the reinforcement bar as shown in Table 13. Subsequently, regardless of the exposure type, the potential for the reinforcement bar to corrode is not significantly related to carbonation and/or moisture. This suggests that there are possibilities for existing chloride within the structure to diffuse and caused reinforcement corrosion.

Table 10. Cross-Tabulation Half-Cell Result * Exposure Type.

		Exposure Type		Total
		Internal	External	
	5%	6	0	6
Half-Cell Potential	50%	0	2	2
	95%	2	0	2
Total		8	2	10

4.1.4. Chloride Content and In-Situ Carbonation Test.

High Chloride content was detected at one of the columns at the first floor and the remaining two at the second floor as shown in Figure 5. The maximum permitted chloride content is 0.40% and the highest detected value is 0.72%, followed by 0.64% as tabulated in Table 12. The mean for overall is around 0.29%, however, the mean for individual floors as per Table 11. The mean for ground floor is 0.18%, the first floor – 0.26% and the second floor – 0.56%. The median for the ground floor is around 0.20%, first floor – 0.17% and the second floor – 0.56%. Although first floor has a very low median for chloride content, it shows that most of its population is in the range between the median to 0.43%. This indicates the high possibility of chloride content in structural members in both the first floor and the second floor.

Table 11. Descriptive Statistics for Chloride Location and Chloride Content.

Chloride Locations	Mean	N	Std. Deviation
Ground Floor	.1800	4	.07659
First Floor	.2600	4	.26432
Second Floor	.5600	2	.22627
Total	.2880	10	.23001

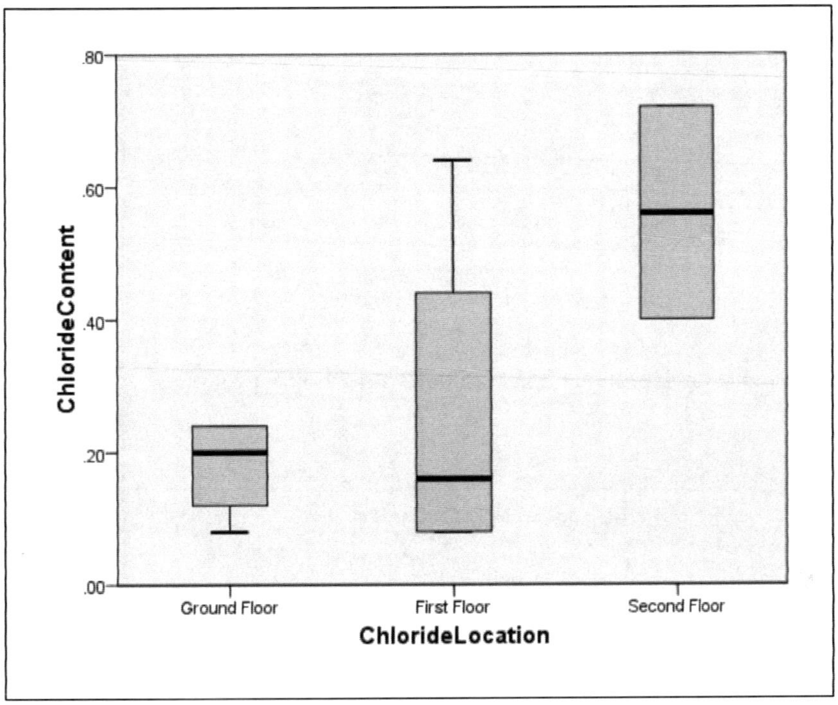

Figure 5. Boxplot - Chloride content (in percentage) vs. Location

Table 12. Descriptive Statistics for Chloride Content.

	N	Range	Min	Max	Mean	SD	Variance	Skewness		Kurtosis	
	Statistic	Statistic	Statistic	Statistic	Statistic	Statistic	Statistic	Statistic	Std. Error	Statistic	Std. Error
Chloride Content	10	.64	.08	.72	.2880	.23001	.053	1.098	.687	.046	1.334

The carbonation depth population dispersion is quite balance with 50% of the sample is above 30 mm with the ground floor exceeding 35 mm and the remainder slightly below 30 mm (Figure 6 & 7). Although the screed at the ground floor is thicker than the rest, this does not hinder the carbonation penetration into the concrete.

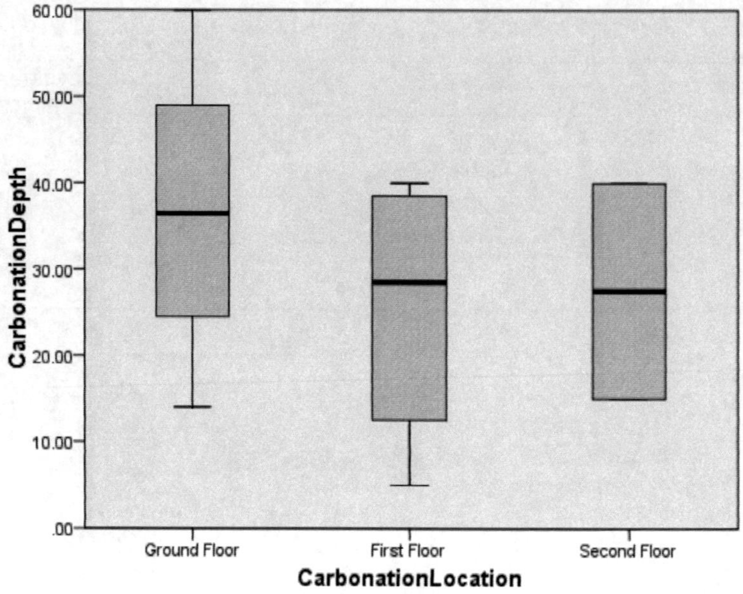

Figure 6. Frequency Histogram and Normal Distribution for Carbonation Depth (in mm).

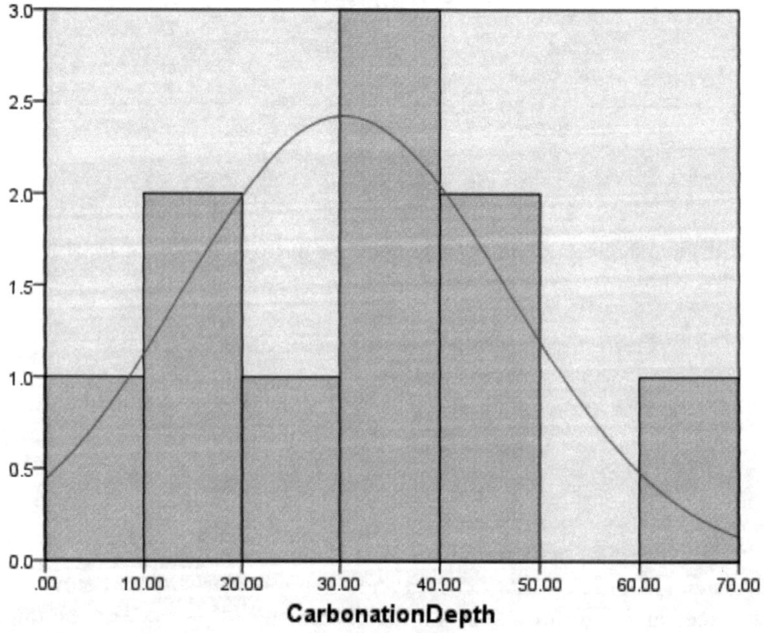

Figure 7. Boxplot -Carbonation Depth (in mm) vs. Location.

4.2. Multivariate Statistical Analysis Using Other Concrete Property Tests

After concluding descriptive analysis or univariate as well as bivariate statistical analysis through cross-tabulation, further relationships and interrelation are required in order to validate if these are significant or insignificant.

4.2.1. Relationship Between Exposure Type, Concrete Cover, Bar Condition and Defective Structure.

The relationship between these data is important in order to understand how concrete cover behaves based on exposure type and reinforcement bars are protected by covers based.

Based on the Multivariate Correlation Analysis in Table 13, a possible linear relationship between exposure type and defective of the structure can be established. This indicate that exposure type is quite related to defective structures found through visual inspection, but it is not significant. The maximum cover shows that there is a possible inverted linear relationship with the bar condition where the thicker the concrete cover, the reinforcement bars would be preserved well. When correlations were made among exposure type, half-cell potential and defective structure; there is no significant relating half-cell potential to other parameters as shown in Table 14. The impact of carbonation depth and chloride content are not showing any significant relationship with exposure type, screed thickness and structural defects as shown in Table 15.

Table 13. Correlation Among Exposure Type, Cover Thickness, Bar Condition and State of Defective.

		Exposure Type	Cover Minimum	Cover Maximum	Bar Condition	Defective Structure
Exposure Type	Pearson Cor	1	.359	-.380	.089	.524
	Sig. (1-tailed)		.154	.176	.403	.060
Cover Minimum	Pearson Cor	.359	1	.216	-.221	.276
	Sig. (1-tailed)	.154		.304	.270	.220
Cover Maximum	Pearson Cor	-.380	.216	1	-.554	-.380
	Sig. (1-tailed)	.176	.304		.077	.176
Bar Condition	Pearson Cor	.089	-.221	-.554	1	.089
	Sig. (1-tailed)	.403	.270	.077		.403
Defective Structure	Pearson Cor	.524	.276	-.380	.089	1
	Sig. (1-tailed)	.060	.220	.176	.403	

Table 14. Correlation – Exposure Type, Half-Cell Potential and State of Defective.

		Exposure Type	Half-Cell Potential	Defective Structure
Exposure Type	Pearson Correlation	1	.250	.500
	Sig. (2-tailed)		.486	.141
	Sum of Squares and Cross-products	1.600	.800	1.000
	Covariance	.178	.089	.111
Half-Cell Potential	Pearson Correlation	.250	1	.250
	Sig. (2-tailed)	.486		.486
	Sum of Squares and Cross-products	.800	6.400	1.000
	Covariance	.089	.711	.111
	Pearson Correlation	.500	.250	1

	Sig. (2-tailed)	.141	.486	
Defective Structure	Sum of Squares and Cross-products	1.000	1.000	2.500
	Covariance	.111	.111	.278

Table 15. Correlation – Exposure Type, Carbonation Depth, Screed Thickness, Chloride Content and Defective State of Structure.

		Exposure Type	Carbonation Depth	Screed Thickness	Chloride Content	Defective Structure
Exposure Type	Pearson Correlation	1	-.185	.242	-.294	.598
	Sig. (2-tailed)		.633	.530	.443	.089
	Sum of Squares and Cross-products	1.556	-10.778	2.667	-.249	1.111
	Covariance	.194	-1.347	.333	-.031	.139
Carbonation Depth	Pearson Correlation	-.185	1	.142	.332	.021
	Sig. (2-tailed)	.633		.715	.383	.958
	Sum of Squares and Cross-products	-10.778	2178.889	58.667	10.524	1.444
	Covariance	-1.347	272.361	7.333	1.316	.181
Screed Thickness	Pearson Correlation	.242	.142	1	-.044	-.506
	Sig. (2-tailed)	.530	.715		.910	.164
	Sum of Squares and Cross-products	2.667	58.667	78.000	-.267	-6.667
	Covariance	.333	7.333	9.750	-.033	-.833
Chloride Content	Pearson Correlation	-.294	.332	-.044	1	-.175
	Sig. (2-tailed)	.443	.383	.910		.652
	Sum of Squares and Cross-products	-.249	10.524	-.267	.462	-.178
	Covariance	-.031	1.316	-.033	.058	-.022
Defective Structure	Pearson Correlation	.598	.021	-.506	-.175	1
	Sig. (2-tailed)	.089	.958	.164	.652	
	Sum of Squares and Cross-products	1.111	1.444	-6.667	-.178	2.222
	Covariance	.139	.181	-.833	-.022	.278

4.2.2. Relationship Between Exposure Type, Chloride Content, Carbonation Depth and Defective Structure.

Since there are unexplained and/or insignificant relationships of these parameters, the available dataset especially on the outcome of defectiveness of structural members against exposure type, carbonation depth, chloride content and exposure type. When the multivariate heat map was produced, evidences redirect the investigation of the concentration of both carbonation depth and chloride content readings based on structural members and locations of the samples as shown in Figure 8 and 9. Beams should be treated for the loss of passivity at the ground floor. Chloride content requires review in making prognosis which is concentrated on the second-floor columns. This shows that different location and structural members require a different approach in repair types.

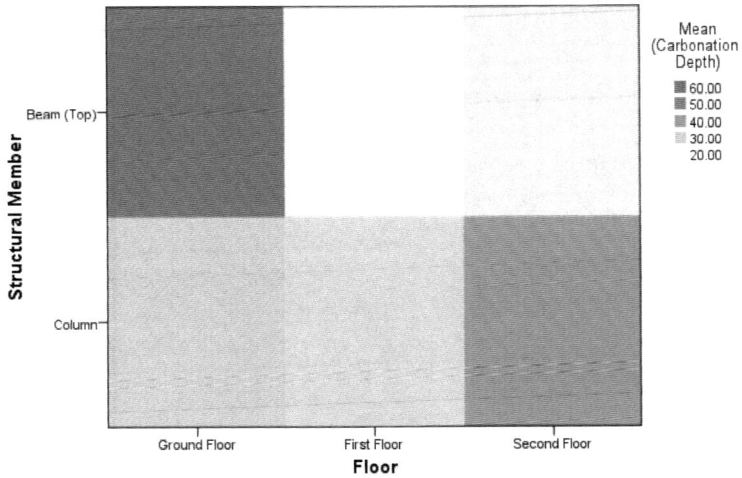

Figure 8. Heat Map - Carbonation Depth, Structural Member and Floor.

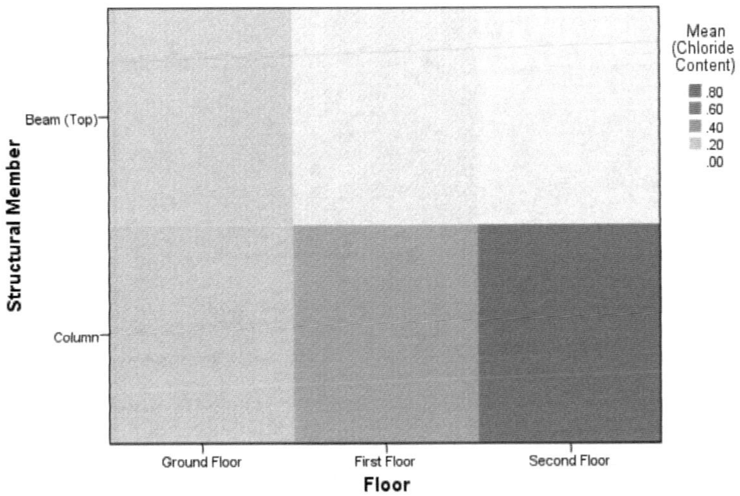

Figure 9. Heat Map - Chloride Content, Structural Member and Floor.

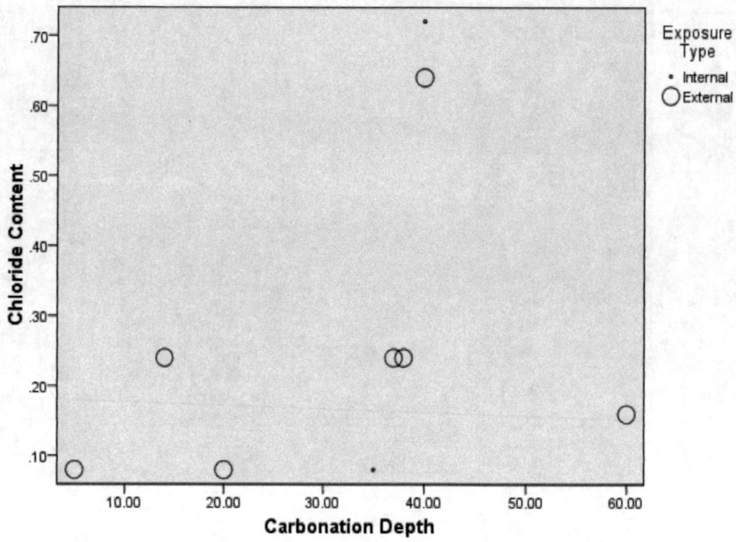

Figure 10. Bubble plot – Chloride Content, Carbonation Depth and Exposure Type.

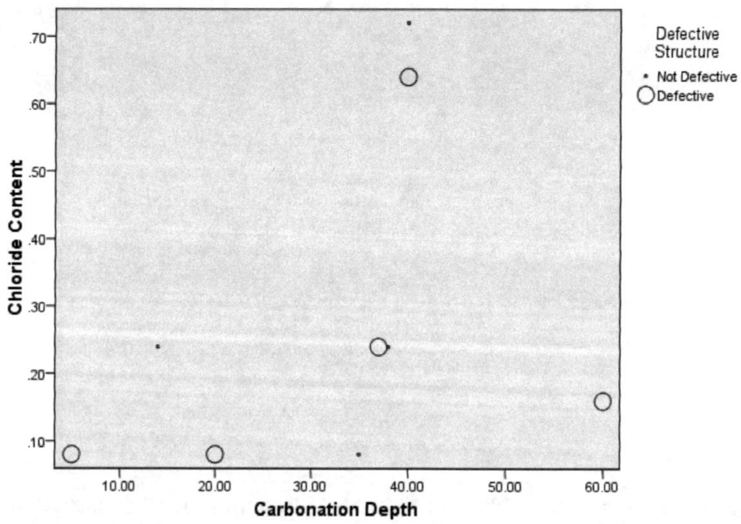

Figure 11. Bubble Plot – Chloride Content, Carbonation Depth and Structural Defectiveness.

Figure 10 indicates that internal and external members are susceptible to chloride attack although the carbonation depth may not be substantial. On the other hand, one key issue which needs to be addressed here is that defects found at the site may not indicate an accurate chemical reaction in the reinforced concrete structural member. Chloride may exist in high concentration without displaying symptoms of degradation or dilapidation. In one instance where two values which have identical chloride content and carbonation depth value resulted in one with proof of degradation and the other one without any indication of possible defects as anticipated in visual inspection.

There are high readings of chloride content on the second-floor columns and a high reading of carbonation depth on the first-floor beam. Detailed analyses are required and boxplots are generated for these parameters.

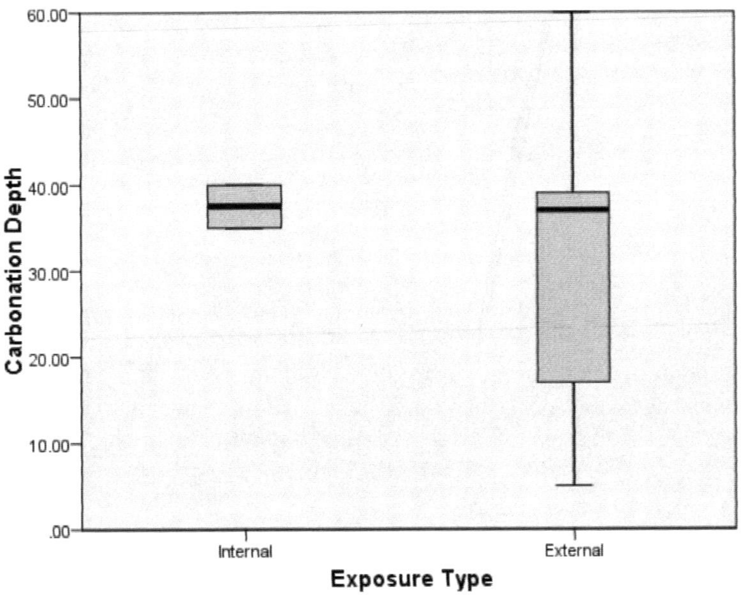

Figure 12. Boxplot- Carbonation Depth (in mm) Vs. Exposure Type.

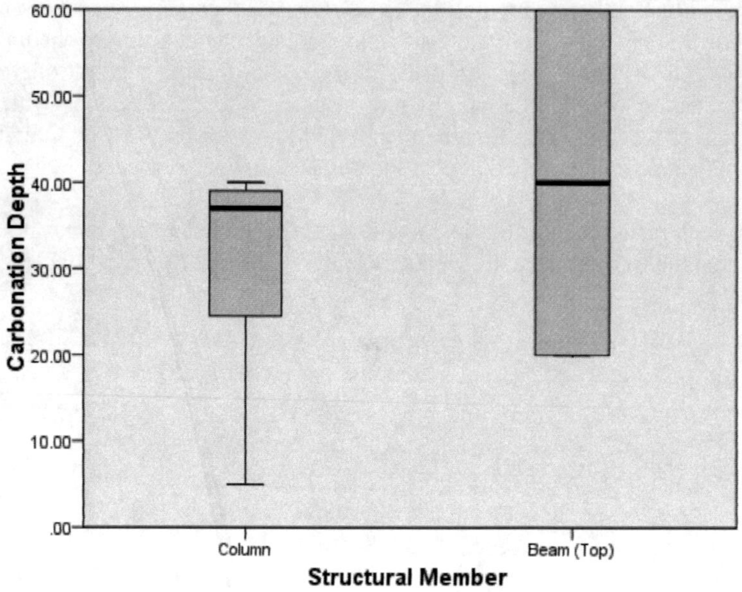

Figure 13. Boxplot- Carbonation depth (in mm) vs. Structural Member.

Figure 13 shows that the average carbonation depth for both column and beam is quite similar. Nevertheless, columns have quite consistent range between 24 mm to 39 mm and beams range between 20 mm to 60 mm. Beams are susceptible to degrading of passivity within the concrete infill.

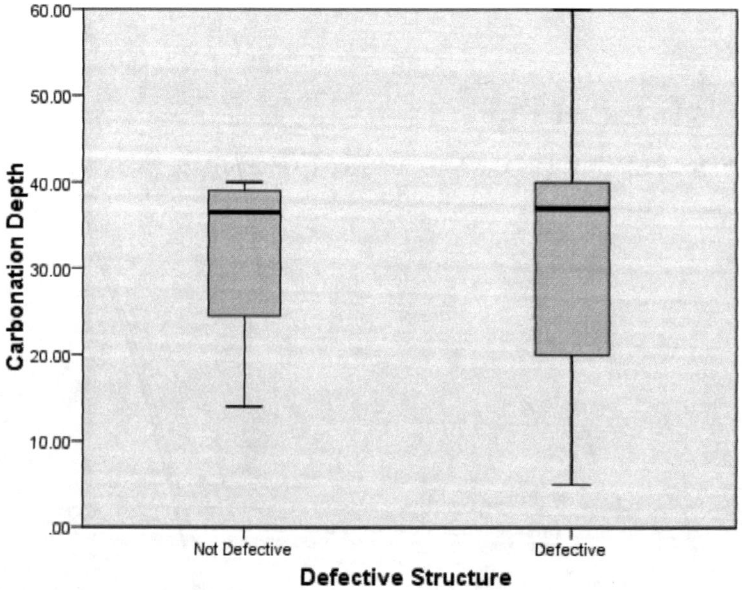

Figure 14. Boxplot- Carbonation depth (in mm) vs. Structural Defectiveness

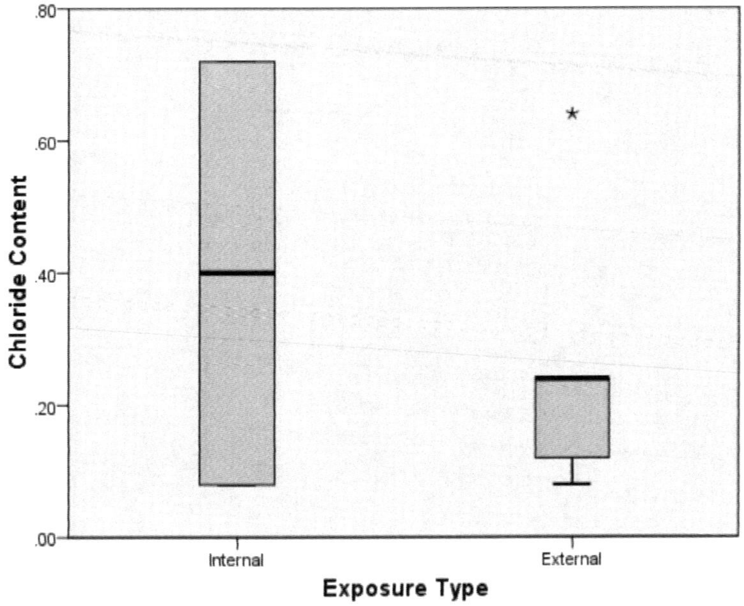

Figure 15. Boxplot- Chloride Content (in percentage) vs. Exposure Type

Figure 15 produced for chloride content show anomalies with one high reading for external structural members which is highly defective and consistent range for internal structural members. Most of the high readings are for columns and chloride content can be high and exist without displaying defects or degradations which are not able to be noticed when conducting visual inspection. Figure 16 indicates that the chloride content is relatively high on columns in comparison with beams. Figure 17 summarized that chloride content cannot be determined through visual inspection without material testing, however, there is high chance that loss of passivity may lead to concrete defectiveness which is traceable through visual inspection.

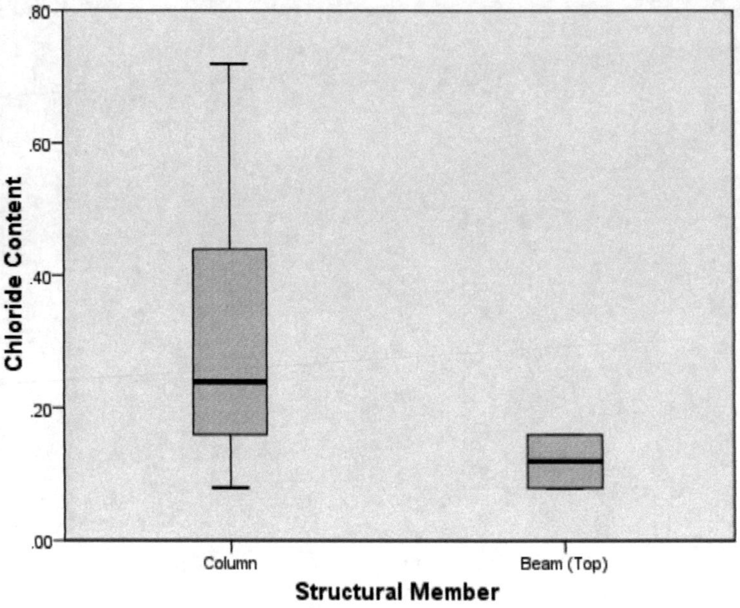

Figure 16. Boxplot- Chloride Content (in percentage) vs. Structural Member.

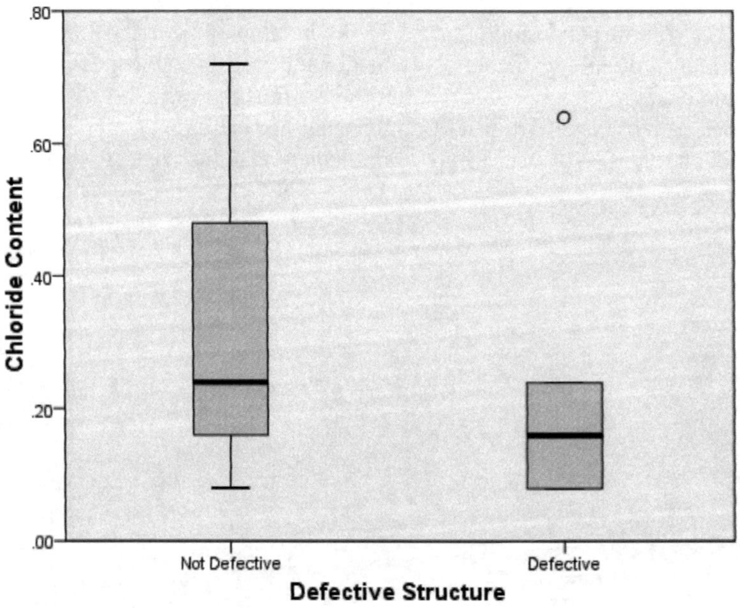

Figure 17. Boxplot- Chloride Content (in percentage) vs. Defective Structure.

4.3. Petrographic Test

Due to statistical analyses which shows another factor which was not captured in the material testing, petrographic examinations were conducted. The petrographic examination shaded the most important and missing relationship which is not captured in this particular in-situ testing. The finding as per report summarized the sample finding as having a low sign of weathering, no crack, no sign of secondary reaction from ASR and AAR.

Nevertheless, it is clear that concrete within this premise has a very high free Cement-Water ratio of 0.65 which signifies the issue in regards to porosity which directly promotes acidic condition due to carbonation and chloride-induced corrosion to reinforcement bars.

5. Result and Discussion

Initial descriptive analysis highlights the value and range for concrete compressive strength and the correlation acquired from rebound hammer test. This prescribed technique complies with the recommended codes of practice BS EN 13791 and MS 1242. However, this does not imply any other reason of why or what led to the degradation of concrete and the lowering of concrete compressive strength.

The cross-tabulation which utilized bivariate statistical analysis managed to improve visualization of data when segregated based on specific groups mainly. The third analyses used in this particular study is the multivariate statistical analysis which tries to establish the correlation of all parameters attained from the site. None of the value provided have significance in confirmatory analysis but there are a few which may have a certain direct or inverted relationship. Maximum cover plays an important role where it shows that it has an inverted relationship with reinforcement bar condition. Thicker concrete cover provides proper protection to the reinforcement bars. The rest indicates that the relationships are in disarray.

Petrographic examination finally revealed the crux of problems for this particular school which was not explainable through all material testing which were conducted. It was the high cement-water ratio during the construction phase which leads to the problem from the concrete cover until the concrete infill of most structural members.

The high chloride content and the extreme carbonation depth which do not produce any defect and not visible to the eye is a concern when these can be overlooked by inspectors when building condition assessment is based on visual inspection based on BCA Guideline by JKR. This construed to the fact these issues were not mitigated and remedied in previous structural repair works.

After diagnosing through material tests and multivariate statistical analyses were conducted, the best solution for structural rehabilitation and repair based on 10 parts BS EN 1504-9:2008, Products and systems for protection and repair of concrete structures – Definitions, requirements, quality control and evaluation of conformity - General principles for use of products and systems. Followings are prognosis suggested based on this study and principles and/or combinations of principles for structure repair and rehabilitation works from BS EN 1504; -

- P2, Moisture control of concrete
- P3, Replacement of damaged concrete
- P4, Strengthening of building components
- P7, Restoring reinforcement passivity
- P10, Cathodic Protection of Reinforcement

6. Recommendation and Conclusion

The result of this study shows that BCA system produced by JKR should not limit testing to buildings which are deemed to be dilapidated from visual inspection. This is proven through multivariate analysis where defects which are visually traceable are not conclusive and have no significance in determining the robustness of the structure. Material testing is essential in the initial stage as a method for piloting the whole building condition assessment. The restriction of only using visual inspection based on tell-tale signs will limit the scope involving building pathology.

After material tests were conducted, it is recommended that data should not be limited to descriptive analysis although this is sanctioned and prescribe by standard specifications. In order to provide proper visualization during diagnosis, the use of multivariate statistical analysis is best in performing diagnostic check not for characteristic strength but also other concrete properties which is essential in providing visualizing the root cause of dilapidations of a building or structure.

When multivariate statistical analysis has no conclusive or significant result based on inferential checks, it is recommended that petrographic examination as the best tool for retrieving missing information which is required in order to provide some insights which can be a basis for holistic prognosis.

7. References

[1] Government Allocates over RM1 billion to Upgrade Dilapidated Schools in Sabah. (2020, July 16). Bernama.

[2] Over 50% of Schools in Sabah Dilapidated. (2018, October 22). Borneo Post.

[3] Noor, S.M., Richard, H.K., Ibrahim, I.S., Sarbini, S.S., Lee, H., & Kumar, J. (2019). *Building Condition Assessment (BCA) on School Building in Sabah, Malaysia. ICOBUILD & RCCE 2019: Material Science and Engineering. Langkawi, Malaysia.*

[4] Jabatan Kerja Raya. (2013). *Garis Panduan Pemeriksaan dan Penilaian Keadaan Bangunan Sedia Ada.*

[5] Watt, D.S. (2008). *Building Pathology: Principles and Practice (2md Ed.).* Backwell Publishing.

[6] Descriptive statistics vs. Inferential Statistics for Research Purpose. (n.d.). World Sustainable. https://worldsustainable.org/descriptive-vs-inferential-statistics/

IOP Conf. Series: Materials Science and Engineering 1229 (2022) 012004 doi:10.1088/1757-899X/1229/1/012004

Monitoring strength development of concrete using the wave propagation technique: A practical study

Z S Tang[1,*], Y Y Lim[1], S T Smith[2], M M H Choi[1] and A Mostafa[1,*]

[1] Faculty of Science and Engineering, Southern Cross University, East Lismore 2480, Australia.
[2] School of Civil, Environmental and Mining Engineering, The University of Adelaide, South Australia, 5005, Australia.

*Corresponding author: z.tang.11@student.scu.edu.au or zisheng87@gmail.com

Abstract. Concrete is one of the most commonly used materials in the construction industry for structural members such as slabs, beams, columns and foundations due to its low cost, high durability and long service life. When concrete is freshly mixed it is in a paste form, which eventually develops into a hard solid once it has cured over time. Concrete typically takes 28 days for full-strength development. In bottom-up concrete construction, it is important to monitor the strength development of concrete under curing to ensure the concrete members are strong enough before casting additional structural elements above. In this study, an attempt is made at a construction site to monitor the strength development of concrete under curing with the non-destructive piezoelectric-based wave propagation (WP) technique. The WP technique employs two piezoelectric sensors and offers the capability of real-time monitoring on the strength development of concrete. The pressure wave (P-wave) velocity can be acquired after applying voltage to the sensor. The P-wave velocity can then be related to the strength of concrete at different curing durations. Based on the on-site experimental study, the P-wave velocities correlate well with the compressive strength of concrete from destructive compressive tests. This on-site experimental study shows the applicability of the WP technique for predicting the strength of concrete in a practical application.

Keywords: Concrete curing, construction site, non-destructive test, wave propagation.

1. Introduction

Concrete is a complex composite construction material consisting of water, cement, fine and coarse aggregates. It undergoes a curing process when the abovementioned raw materials are mixed together. Physically, the freshly-mixed concrete is in a paste form and gradually develops into a hard solid throughout the curing duration. At the same time, concrete starts to develop its strength throughout the curing process. Generally, concrete achieves its full strength after 28 days of curing [1-2]. Therefore, concrete must be sufficiently cured (i.e. achieve adequate compressive) before formwork supporting it can be safely removed. Traditionally, destructive compression tests can practically determine the compressive strength of concrete using separately cast concrete samples. However, concrete strength determination from such destructive test cannot fully represent the concrete in place.

The emergence of smart materials offers a promising solution in structural health monitoring (SHM) for monitoring the concrete strength throughout the curing process [3]. One of the most commonly used

IOP Conf. Series: Materials Science and Engineering 1229 (2022) 012004 doi:10.1088/1757-899X/1229/1/012004

smart materials in SHM is a piezoelectric material that is made from lead zirconate titanate (PZT), and it is in the form of patches. With piezoelectricity, the PZT patch can act as a sensor or an actuator [4]. When a PZT patch is applied with an electrical field, it undergoes a mechanical strain through the converse effect. At the same time, it generates electricity when the PZT transducer is subjected to mechanical stress through the direct effect. For so, the PZT patches can be surface-attached onto or embedded into the host structure (i.e. the material to be monitored) to determine the mechanical properties, including strength, elastic modulus and Poisson ratio [5]. Due to the high sensitivity of the PZT patches, they are capable of monitoring the changes of compressive strength of concrete throughout the curing duration in a real-time manner without incurring damage on the concrete.

In the past, one of the piezoelectric-based techniques, namely the wave propagation (WP) technique, was employed for cure monitoring of concrete by measuring elastic wave velocity changes in concrete at various curing duration. Some previous studies have shown the capability of the WP technique for monitoring the early concrete strength development of concrete [6-8]. Lim et al [9] also showed that the WP technique is capable of determining different concrete strengths by measuring the wave velocities travel through the concrete at various curing durations, despite the specimens having a similar water-cement ratio. Additionally, Lim et al [9] conducted a long-term monitoring study on the concrete curing process.

Although the previous studies have proven the WP technique's capability for monitoring the curing process of concrete in laboratory, there is no known study demonstrating its performance at the construction site. Therefore, the WP technique is adopted in this study to monitor the concrete curing process by measuring the wave velocity for determining the compressive strength of concrete under curing.

2. Experimentation

The experimental study was conducted at a construction site in Malaysia. In particular, this study was conducted on the third floor of one of the 20-storey high rise residential buildings under construction. A pour strip (i.e. a section of a slab) was selected for the experimental study due to the convenience and accessibility to the site.

2.1. On-site concreting

The concrete of the slab is Grade 35 normal-mix concrete. The mix proportions of concrete are provided by the batching plant, and they are tabulated in Table 1. The freshly mixed concrete was delivered to the construction site from batching plant with the concrete truck, as shown in Figure 1(a) and transferred to the testing location using a tower crane as shown in Figure 1(b). Since this is an in-situ monitoring study, the concrete is air-cured.

Table 1. Mix proportions of concrete provided by batching plant.

Characteristic strength (N/mm^2)	35
Target mean strength (N/mm^2)	42.5
Slump (mm)	75± 25
w/c ratio	0.49
Ordinary Portland Cement (kg)	350
Coarse aggregate – crushed granite (kg)	930
Class C fine aggregate – natural sand (kg)	878
Water (kg)	170

(a) (b)

Figure 1. Concreting at construction site: (a) transfer of concrete to skip and (b) transfer of concrete to testing location.

In order to acquire the changes of compressive strength at various curing durations, the piezoelectric-based WP technique was employed. The details of the experimental setup are discussed further in the following section

2.2. Experimental setup at construction site

The current study employed mortar-based cylindrical SA for acquiring signatures using the WP technique. Figure 2(a) shows the schematic of an SA, where a waterproof PZT patch was embedded in the freshly mixed mortar. Upon curing for a minimum of 28 days, the SAs were placed, at a spacing of 80 mm, into freshly mixed concrete, as illustrated in Figures 2(b) and 2(c).

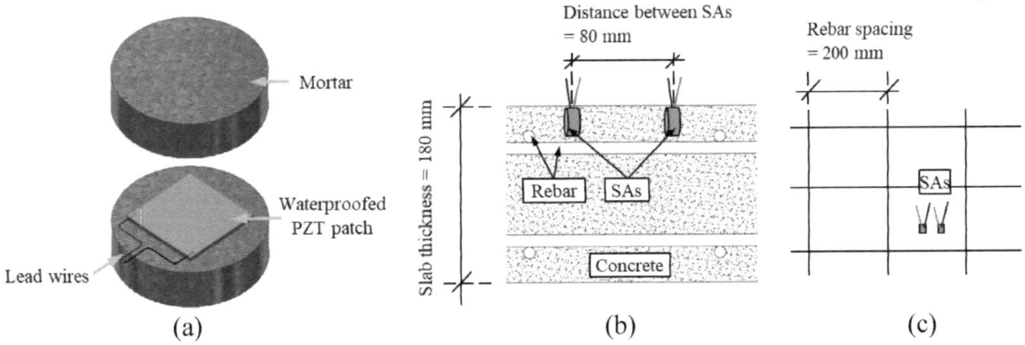

Figure 2. Smart Aggregate (SA): (a) Cross section of SA, (b) side view of embedded SAs in concrete slab and, (c) top view of installation of SAs

The WP technique's setup consists of a function generator, an oscilloscope, an amplifier and a computer, which was transferred to the construction site for acquiring signatures, as shown in Figure 3. One of the SAs acts as an actuator, and the other one acts as a sensor [10]. For this study, a transient actuating voltage at 100 kHz was generated by the function generator and amplified by the amplifier before applying it across the polarisation direction of the actuator. The applied electricity would then induce mechanical vibration and form an elastic wave to propagate radially across the concrete. The sensor then reverted the elastic wave into sensing voltage. Both the actuating and sensing voltages were captured by the oscilloscope and recorded using a computer. Signatures were acquired right after the placement of the SAs, and the whole monitoring process took 28 days.

Figure 3. On-site experimental setup: (a) WP technique, and (b) embedded SAs in concrete as in the solid circles.

3. Results and discussions

3.1. Non-destructive piezoelectric-based WP test

A typical WP-acquired signature as shown in Figure 4 consists of an actuating voltage and a sensing voltage. There are several types of waves that can be identified from the WP technique, such as pressure wave (P-wave), shear wave, Rayleigh wave and others [11-13]. In general, P-wave is the most suitable to be employed for monitoring the curing process of concrete [10]. This is because it is a non-dispersive wave and it has the highest propagating speed [14-15]. From Figure, P-wave packet is reflected from the first wave packet of the sensing voltage. The velocity of the wave is evaluated by dividing the distance between the actuator and the sensor by the travelling duration of the wave, that is the time of flight (TOF). According to Figure 4, the TOF of the P-wave can be determined from the time difference between the peak of actuating voltage and the first peak of the sensing voltage.

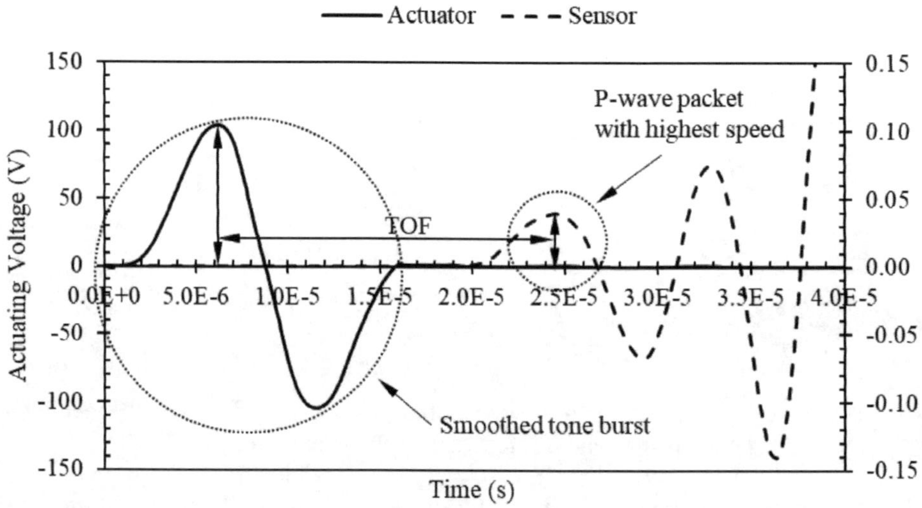

Figure 4. Actuating and sensing voltages of concrete after seven days of curing at 100 kHz actuating frequency.

Changes in P-wave velocity throughout the curing duration are summarised in Figure 5. The wave velocity increases rapidly for the first day of curing, showing a rapid strength development of concrete strength. The rate of increase of wave velocity decreases gradually, indicating the strength of concrete is moderately developed until seven days. After that, the wave velocity only slightly increases until the 28-day mark.

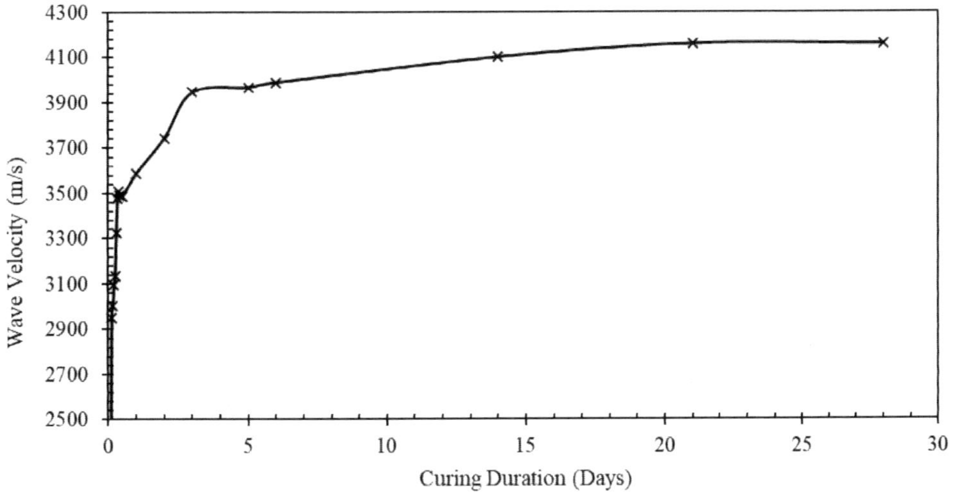

Figure 5. WP-acquired wave velocity at various curing durations.

3.2. Destructive compressive strength test

The compressive strength of concrete over 28 days of curing durations as determined from destructive compressive strength tests is presented in Figure 6. The compression test was done on the cube specimens of 150 mm in accordance to BS EN 12390-3. Concrete underwent rapid strength development for the first day of curing and achieved a compressive strength of 17.5 MPa. The concrete strength developed further to 32 MPa at the 7-day mark and achieved a value of 39 MPa at day 28.

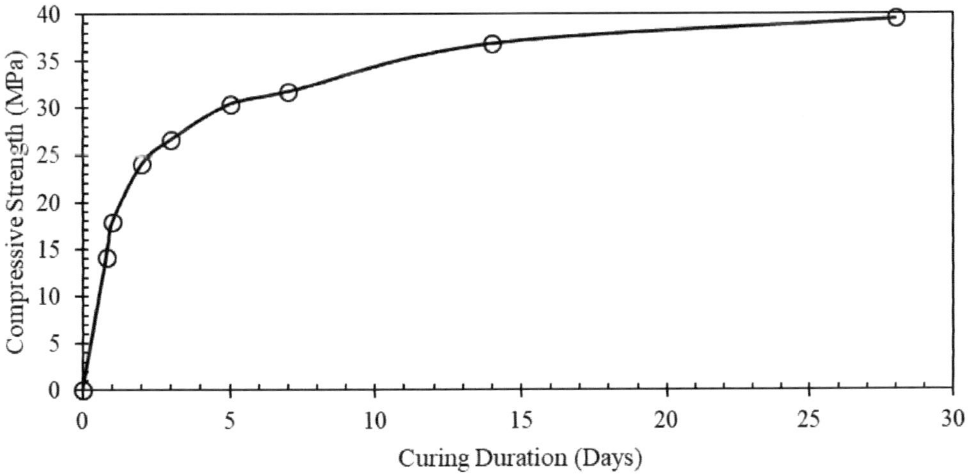

Figure 6. WP-acquired wave velocity at various curing durations.

As from Figures 5 and 6, the changes in wave velocity and compressive strength of concrete over 28 days are similar. Therefore, a correlation between two parameters can be developed, which is presented in the following subsection.

3.3. Correlation of WP-acquired wave velocity and compressive strength

The relationship between compressive strength and wave velocity is presented in Figure 7. Based on the current study, an exponential empirical relationship was identified through the curve fitting. The correlation shows strong exponential relations between compressive strength and wave velocity, with a coefficient of determination (R^2) of 0.961. The empirical equation is expressed in Equation (1).

$$f_c' = 0.3894e^{0.001098v(t)} \tag{1}$$

where f_c' = concrete compressive strength at a specific curing duration using the WP technique (MPa) and $v(t)$ = wave velocity at a specific curing duration (m/s) in a range of $3000 \leq v(t) \leq 4400$.

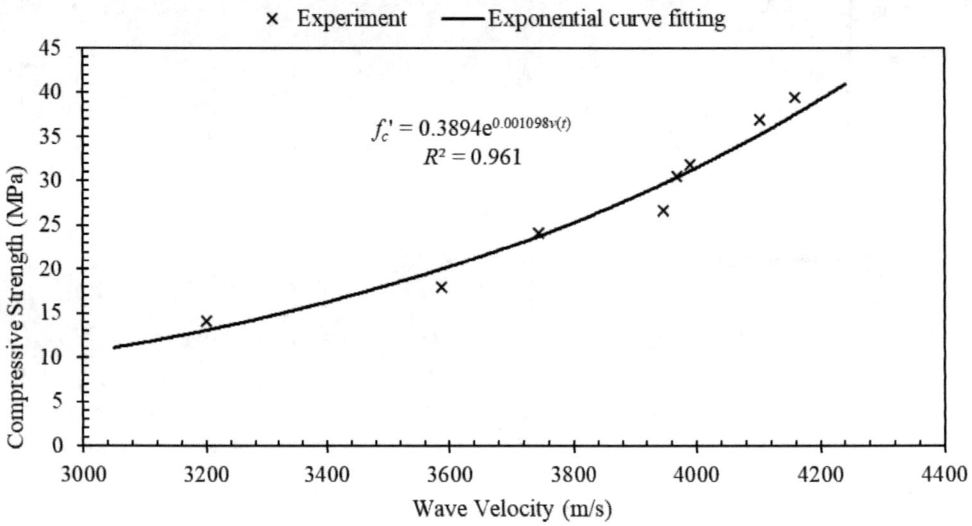

Figure 7. WP-acquired wave velocity at various curing durations.

4. Conclusions

In this study, an on-site monitoring study of concrete curing is successfully carried out using the piezoelectric-based monitoring technique. Based on the study, changes in wave velocity showed a good representation of the strength development of the on-site concrete throughout the curing process. An exponential equation was developed to determine the strength of concrete from the wave velocity. This study can be further extended to predict the stiffness of concrete using the wave equation. Additionally, the outcomes of the study can also be verified with the traditional ultrasonic pulse velocity (UPV) test, as well as the EMI technique.

5. References

[1] Zhang C, Panda G P, Yan Q and Zhang W 2020 Construction and Building Materials 258 120348
[2] Sharma S and Mukherjee A 2015 Cement and Concrete Composites 55 337-347[3]
[3] Lim Y Y, Smith S T and Soh C K 2018 NDT & E International 99 50-63
[4] Lim Y Y, Tang Z S and Smith S T 2019 Smart Materials and Structures 28(1) 015016

[5] Tang Z S, Lim Y Y, Smith S T and Izadgoshasb I 2019 Mechanical Systems and Signal Processing 128 172-190
[6] Dumoulin C, Karaiskos G, Carette J, Staquet S and Deraemaeker A 2012 Smart Materials and Structures 21(4) 047001
[7] Kong Q, Hou S, Ji Q, Mo Y L and Song G 2013 Smart Materials and Structures 22(8) 085025
[8] Lim Y Y, Kwong K Z, Liew W Y H and Soh C K 2016 Smart Materials and Structures 25(8) 085021
[9] Lim Y Y, Kwong K Z, Liew W Y H and Soh C K 2017 Construction and Building Materials 152 506-19
[10] Tang Z S, Lim Y Y, Smith S T and Izadgoshasb I 2019 Mechanical Systems and Signal Processing 128 172-190
[11] Song F, Huang G, Kim K and Haran S 2008 Smart Materials and Structures 17(5) 055024
[12] Mohseni H and Ng C T 2018 Structural Health Monitoring 18(1) 303-317
[13] Yu H, Lu L and Qiao P 2020 Construction and Building Materials 242 118033
[14] Gupta S, Fan G and Loh K J 2017 IEEE Sensors Letters 1(5) 1-4
[15] Wojtczak E, and Rucka, M 2021 Mechanical Systems and Signal Processing 151 107397

Acknowledgements
The authors would like to acknowledge Juteras Vision Sdn Bhd for offering an opportunity to undertake the experimental study at a construction site.

A study on monitoring the hydration process of glasscrete using electromechanical impedance technique

M H M Choi[1,*], Z S Tang[1], R Vasquez Padilla[1], Y Y Lim[1] and A Mostafa[1]

[1] Faculty of Science & Engineering, Southern Cross University, Lismore 2480, Australia.

*Corresponding author: m.choi.16@student.scu.edu.au

Abstract. Glass has been suggested as a viable material to replace aggregates in concrete to maintain sustainable development for the future. However, a concern is raised about the cementitious reaction with glass that could cause concrete spalling and loss of strength. Therefore, monitoring the concrete hydration process and the structural health throughout its service life is essential. In this study, the hydration process of lab-scale glasscrete prisms is monitored using the electromechanical impedance (EMI) technique. This technique employs one piezoelectric-based transducer that encases a material called Lead Zirconate Titanate (PZT) to operate as an actuator and senor. The experimental result shows that the embedded transducer is an effective means to monitor the glasscrete hydration process. The acquired admittance signature was able to capture the strength development during the glasscrete hydration period via interpreting the PZT and structural resonance peaks

Keywords: Lead Zirconate Titanate (PZT), electromechanical impedance technique, smart materials, concrete hydration, non-destructive evaluation.

1. Introduction
Concrete is one of the most widely used materials in construction due to its durability and superior fire resistance. However, the production of its composition cement contributed approximately eight per cent of the world's carbon dioxide (CO_2) emission [1]. Another environmental issue with concrete is the excessive dredging of riverbed and seafloor that causes detrimental effects on animals' habitats and nearshore communities [2]. Therefore, scientists and engineers seek alternatives to minimise the use of or even replace cement and sand for concrete to maintain sustainable development for the future, especially during the current construction boom around the world.

One of the viable materials for replacing sand is glass which is widely used in our daily life for packaging, tableware, appliance and automobile. Currently, the waste recovery rate for glass is considerably high at approximate 59% in Australia with a 100% recyclable rate, but recycling the collected glass into new glass containers is often labour- and energy-intensive [3]. It requires the collected glass free from contamination and sorted into the same colour before being remelted for the recycling process. Therefore, the commodity value for recycling glass is low, leading to several studies on the possibilities of mixing recycled glass into concrete. This suggestion can virtually reduce waste in landfills, carbon emissions and save scarce resources. Thus far, few projects have even put this idea into real-world practice as a pavement component [4, 5].

Nonetheless, there is a drawback with glass mixing with cementitious material where the alkali-silica reaction (ASR) could occur, causing concrete spalling and loss of strength in long term settings [11]. Therefore, determining and monitoring the concrete strength development during and after the hydration process is imperative for structural safety. Several techniques are currently available to detect changes in these mechanical properties, such as stiffness and elastic modulus, and they can be categorised as destructive and non-destructive methods. However, these methods all pose some limitations in terms of cost, accuracy, sensitivity, and practicality. For example, the market available non-destructive testing (NDT) method, ultrasonic pulse velocity (UPV), requires placing expensive transducers opposite to each other for accurate results [6].

The piezoelectric-based monitoring technique is one of the NDT methods that might overcome limitations that other techniques have and provide a more efficient, cost-effective, and real-time monitoring capability [7]. One of the most adopted piezoelectric-based monitoring techniques is the electromechanical impedance (EMI) technique. It only requires one piezoelectric-based transducer that encases a material called Lead Zirconate Titanate (PZT) to operate as an actuator and sensor. The fundamental principle of EMI is that the transducer is externally excited to vibrate at a high frequency by an alternating voltage, and the structural response is captured simultaneously.

To date, research related to monitoring plain concrete's hydration process using this technique was found to be very groundbreaking. However, no study has been completed to investigate its effectiveness on concrete mixed with glass, glasscrete [7 - 9]. Thus, it is valuable to investigate the applicability of the piezoelectric-based technique in evaluating the strength development process of glasscrete.

2. Experimental Setup

Two PZT patches, with the dimensions of 15 x 15 x 0.5 mm, are first protected with three layers of two parts epoxy to improve their water resistance. After that, the protected PZT patches are embedded in a mortar to increase their abrasion resistance. The mortar mix has the following properties: water-cement ratio of 0.7 and cement-sand ratio of 0.33. The protected PZT patches, named smart aggregates, will then be cured for 28 days before use. Figure 1(a) - (c) show the three-step procedures of preparing smart aggregates.

(a) (b) (c)

Figure 1. Pictorial representation of showing the procedures of making smart aggregate.

For the glasscrete mix, the characteristic strength of the design mix is 40 MPa, and all fine aggregate is replaced with recycled glass obtained from Lismore Community Recycling Centre, New South Wales, Australia, as shown in figure 2(a). The recycled glass is untreated and used directly into the concrete mix. The glasscrete mix is then poured into two 75 mm x 75 mm x 300 mm length prism moulds, and 15 (Ø100 x 200 mm) cylinders are cast for the compressive strength test. The smart aggregate is embedded before the glasscrete set, and the final product of the prisms are shown in figure 2(b). Finally, the glasscrete is water cured for 28 days after being demolded, 24 hours after cast.

An impedance analyser (Hioki IM3570) is used to obtain the admittance signature from the smart aggregate to monitor the hydration process. A 5V alternating voltage with a wide range of frequency (5kHz – 885kHz) is applied across the smart aggregate to record its electrical admittance.

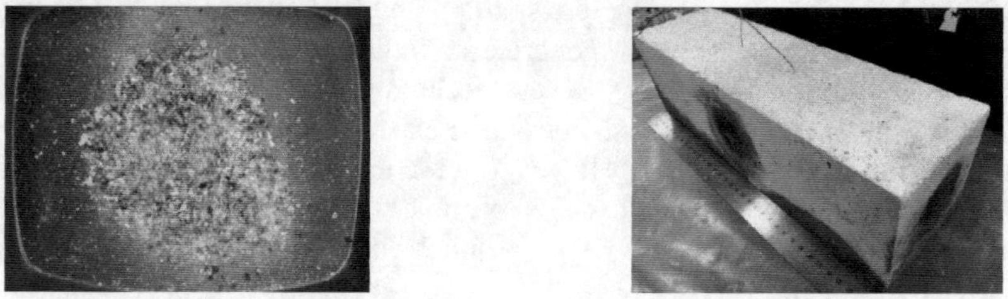

(a) (b)

Figure 2. (a) The recycled glass and (b) the glasscrete.

3. Glasscrete Hydration Monitoring

3.1. PZT Resonance Peak

Figure 3 illustrates the first PZT resonance peak during the 28 days of curing. There is a progressive shifting of the resonance peak to the right, indicating the changes in mechanical properties such as modulus of elasticity and compressive strength. The strength development of the glasscrete is the main cause of shifting the PZT resonance peak as the vibratory transducer only interacts and is constrained by the glasscrete.

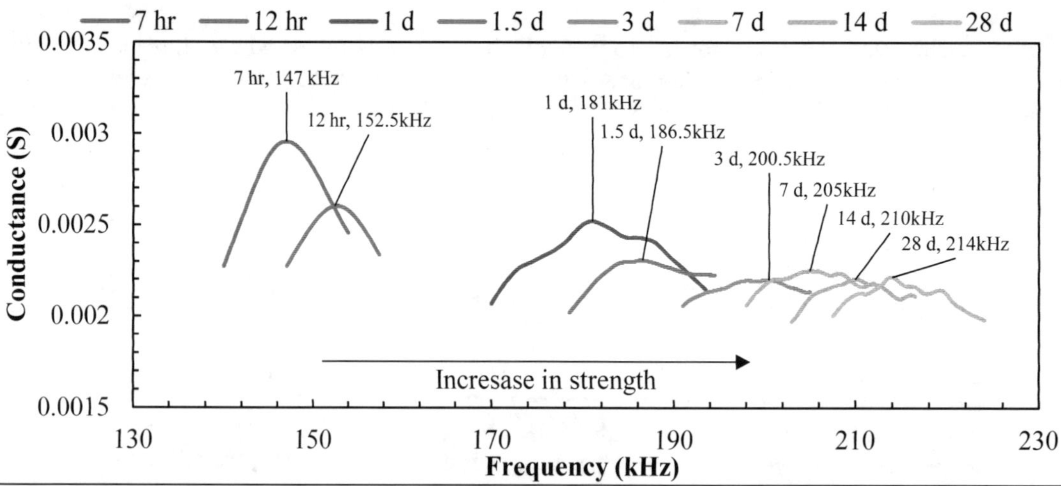

Figure 3. Conductance signature versus frequency acquired from glasscrete in the frequency range of 130 – 230 kHz.

A phenomenon is noted in figure 4 where the rate of change of the first PZT resonance frequency peak correlates well with the rate of change from the traditional compressive strength test. The increase in resonance frequency was high initially, indicating the initial rapid reaction rate during the hydration process in glasscrete curing. The rate then slows down after seven days of curing. The result shows that despite the presence of the glass inside the glasscrete, the EMI technique is still capable of capturing the change in strength as the glasscrete cure.

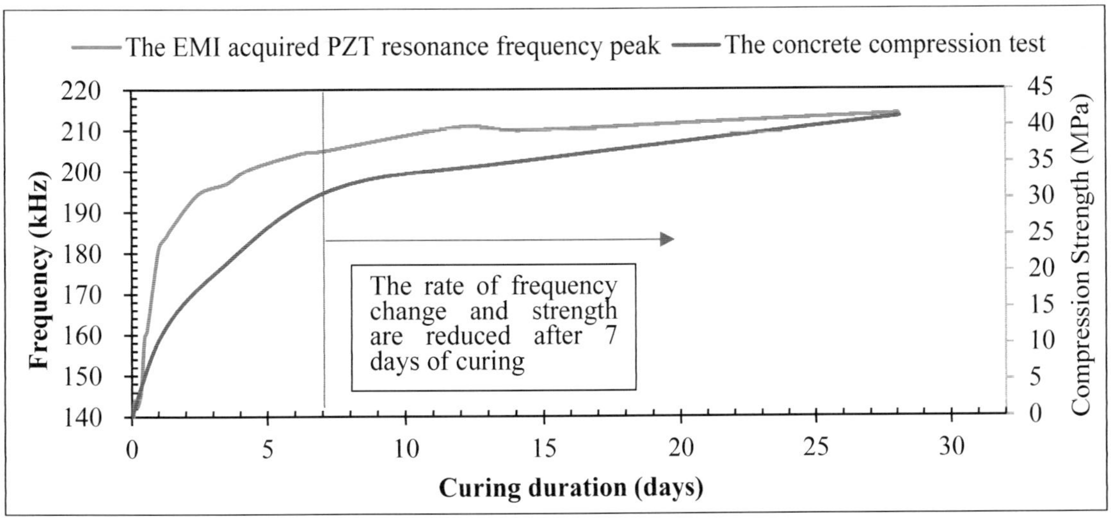

Figure 4. The EMI acquired PZT resonance frequency peak and the glasscrete compressive test result against curing time

3.2. Structural Peak

Lim [10] reported that the structural resonance peaks, which occur below 100 kHz, are more effective in characterising the hydration process after one day of curing. As depicted in figure 5, the structural resonances become more apparent after curing for one day, indicating the glasscrete has gained sufficient strength that the smart aggregate could monitor the glasscrete resonance frequency. In addition, a similar rate of change for the hydration process over 28 days is observed using the structural peak.

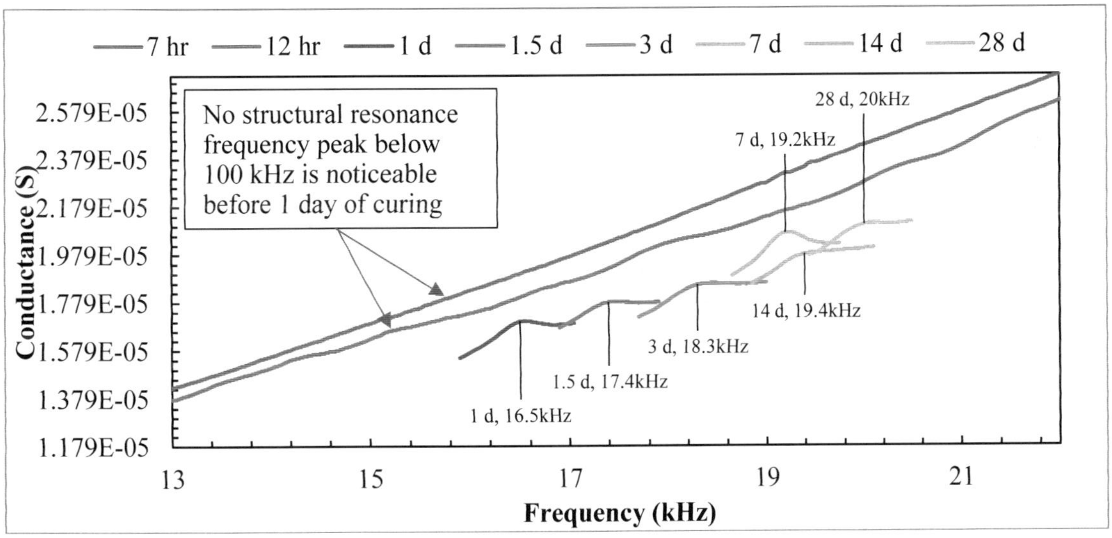

Figure 5. Conductance signature versus frequency acquired from glasscrete in the frequency range of 130 – 220 kHz.

3.3. Relationship between EMI acquired first PZT resonance frequency peak and compressive strength

An empirical relationship between the EMI acquired first PZT resonance frequency peak and compressive strength is expressed in equation (1). As shown in figure 6, a rigorous exponential relationship between these two parameters was identified using the least square method, given a coefficient of determination (R^2) equal to 0.996. Please note that equation (1) is only valid for this experiment due to the manufacturing variance of the smart aggregates.

$$f_c' = 0.0092e^{4\times10^{-5}f(t)} \tag{1}$$

where f_c' = glasscrete compressive strength at a specific curing duration and $f(t)$ = the EMI acquired first PZT resonance frequency peak at a specific curing duration (kHz).

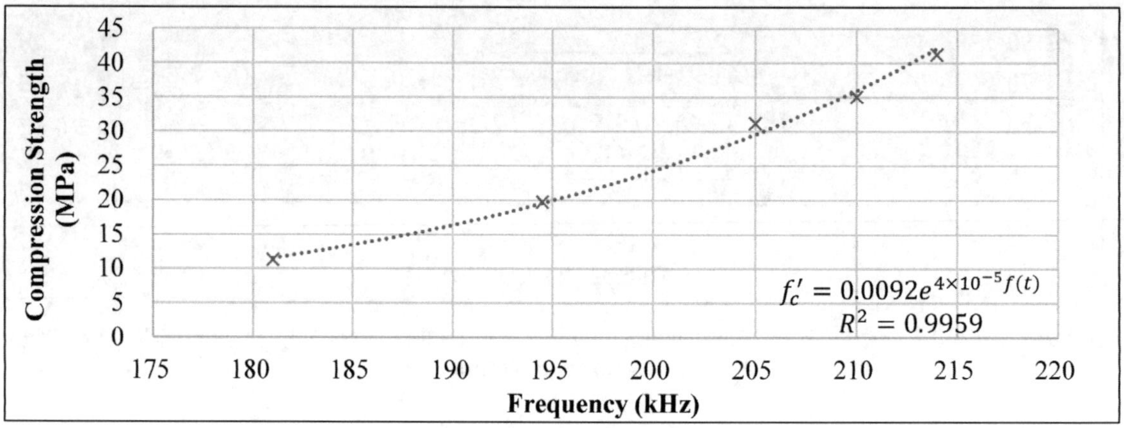

Figure 6. Empirical relationship between EMI acquired first PZT resonance frequency peak and compressive strength

4. Conclusion

In this study, the feasibility of employing the EMI technique to monitor the hydration process of glasscrete is demonstrated by observing the movement of PZT and structural resonance peaks. It found that the acquired admittance signatures have a rightward shift of resonance frequency peak, indicating the strength gain in glasscrete during hydration. Further study is needed to determine the empirical relationship between the admittance signatures and physical properties of glasscrete, such as the elastic modulus and compressive strength. In addition, the high sensitivity of the EMI technique has the potential to detect the alkali-silica reaction (ASR) in glasscrete. However, it would require the experiment to carry on for several months or years before internal cracks are formed to be detected by the EMI technique.

5. References

[1] Climate change: The massive CO2 emitter you may not know about [Internet]. BBC News. 2021 [cited 18 October 2021]. Available from: https://www.bbc.com/news/science-environment-46455844

[2] The world is running out of sand -- and there's a black market for it now [Internet]. Business Insider Australia. 2021 [cited 18 October 2021]. Available from: https://www.businessinsider.com.au/world-running-out-sand-resources-concrete-2018-6

[3] Final National Waste Report 2020 [Internet]. Blue Environment Pty Ltd; 2020. Available from: https://www.awe.gov.au/sites/default/files/env/pages/5a160ae2-d3a9-480e-9344-4eac42ef9001/files/national-waste-report-2020.pdf

[4] Roads G. Hanson Trials Recycled Glass in Concrete [Internet]. Green Roads. 2021 [cited 18 October 2021]. Available from: https://greenroads.com.au/projects/hanson-trials-recycled-glass-in-concrete/

[5] Guo P, Meng W, Nassif H, Gou H, Bao Y. New perspectives on recycling waste glass in manufacturing concrete for sustainable civil infrastructure. Construction and Building Materials. 2020;257:119579.

[6] Guidebook on non-destructive testing of concrete structures. 17th ed. Vienna, Austria: The International Atomic Energy Agency (IAEA); 2002.
[7] Lim Y, Kwong K, Liew W, Soh C. Non-destructive concrete strength evaluation using smart piezoelectric transducer—a comparative study. Smart Materials and Structures. 2016;25(8):085021.
[8] Lim Y, Smith S, Soh C. Wave propagation based monitoring of concrete curing using piezoelectric materials: Review and path forward. NDT & E International. 2018;99:50-63.
[9] Lim Y, Smith S, Padilla R, Soh C. Monitoring of concrete curing using the electromechanical impedance technique: review and path forward. Structural Health Monitoring. 2019;20(2):604-636.
[10] Lim Y. Monitoring of concrete hydration using electromechanical impedance technique. 23rd Australasian Conference on the Mechanics of Structures and Materials (ACMSM23). Byron Bay, Australia: Southern Cross University; 2014. p. 1155-60.
[11] Sun L, Zhu X, Kim M, Zi G. Alkali-silica reaction and strength of concrete with pretreated glass particles as fine aggregates. Construction and Building Materials. 2021;271:121809.

IOP Conf. Series: Materials Science and Engineering 1229 (2022) 012006 doi:10.1088/1757-899X/1229/1/012006

The implementation of inclusive design for public buildings in Kota Kinabalu, Sabah

J Julius[1], L Gungat[1] and I Tukiman[2]

[1]Civil Engineering Programme, Faculty of Engineering, Universiti Malaysia Sabah, Sabah 88400, Malaysia

[2]Department of Landscape Architecture, Kuliyyah of Architecture & Environmental Design, International Islamic University Malaysia, Gombak, Selangor, Malaysia

*lillian@ums.edu.my, izawati@iium.edu.my and jenjulius@gmail.com

Abstract. Inclusive design is a main aspect in designing public buildings. Lack of accessibilities (ramp entrance, task lighting, lever door handles and etc.) as part of inclusive design methodology in public buildings used by the people for any purpose, such as, education, entertainment, worship or assembly has decreased a quality of international standard building design which directly has effected end user's lifestyle, time and money. As the demand to equip our public building with inclusive design facilities has increased due to rapid population growth. Hence, this research paper presents the inclusive design implementation in the existing public buildings in Kota Kinabalu, Sabah which includes identification of design barriers and challenges for persons with disabilities (PWDs), children, elders and caretakers. In summary, this study identified the lack of inclusive design elements for buildings built in the 1980s, where the enactment of the Persons with Disabilities Act 2008 was only enforced on July 7, 2008. Therefore, these buildings improvement will keep people's diversity and uniqueness by experts proper planning based on experience and feedback which can be found in buildings built in the 2009 and later.

1. Introduction

Inclusive design term was first used in 1994 [1] and since that year it was frequently applied, the worldwide implications due to population has made the Inclusive Design as preliminaries focus for disability and ageing in typical design concept and challenges. Opportunities in business to promote Inclusive Design elements did not meet the target in the way that earlier implementation was not mandatory.

Inclusive design is a core in a preliminaries design approaches to designing for persons with disabilities (PWDs), small children, elders and caretakers. as a population subgroup, and as a part of segment in international facilities; the integration trend towards of persons with disabilities (PWDs), small children, elders and caretakers in the ordinary society. This integration trend has demonstrated fundamentally in the multitude numbers based on the local climate, culture, circumstances, social environments and surroundings [1]. Different types of users live together in society.

Primary focus in the US are mostly on how to access the built environment and public places as the independent's right, with reference to the Americans with Disabilities Act 1990 (ADA,1990), and with coming from the civil rights movement solid lead. In the UK primary focus was extended to include access to services, stated in the Disability Discrimination Act 1995 (DDA,1995). The UK Equality Act

2010 (UK, 2010) has updated their previous legislation and the new regulations will be implemented gradually in near future. In the Europe, planners are to include access to information and related services through the internet and communications technology (ICT) media under the aegis of the European Union (COM, 1999).

"Inclusive Design is not a specialist job or new design concept. It is a universal proposition for planners and designers to make sure that their design incorporating the requirements of many variation and possible users, regardless of age or capability. Inclusive Design has been powered by two major trends which is Design for All in Europe and Universal Design in USA – population, ageing and the growing movement to integrate persons with disabilities (PWDs) into ordinary society" (Design Council, 2008).

In design aspect, planners and designers should take into account the population of society as locating as set name of 'bell curves' - with high tails level and low size capability (in the anthropometrics shape) and 'average' capability or size of a central bulge which is mostly will change dependent on the chosen population as shown in Figure 1. In terms of the capability demands which they place on the user, it can consider to put in products, services and the built environment, then it will be possible to show sectors of the population as to set availability according to the level of capability demanded by the product, service or environment [1].

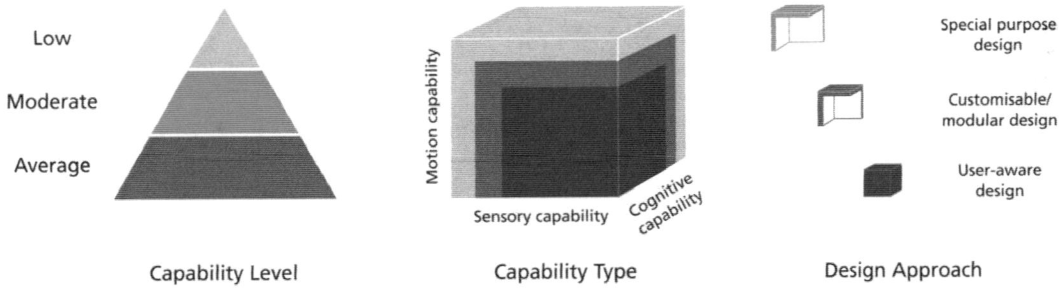

Figure 1. The Inclusive Design Cube

Inclusive Design capable to contribute significant impact on all stages of implementation, the principles and methods can be implemented in other related aspects, such as design for patient safety where the focus is on improving the quality of public sector services. Furthermore, Inclusive Design also can very much upgrade the captivation quality of public spaces and surrounding environments dedicated to international tourism and major international events and attractions. For example, there is a strong prominence requirement on Inclusive Design aspect in the planning and construction of Olympic standard sporting facilities (ODA, 2008a, 2008b).

Three origins of influences and ideas on inclusive design as illustrated in Figure 2 named a concept coined and establish in 1994 in the United Kingdom (UK). It concentrates on design for the general public. Universal design established in the United State of America (USA) which with a major influence on the emergence of universal design in Japan. Design Age, Design for ageing network and European (EU) emerging with other groups. In terms of universal design, Japan's universal design represented the realities of the world's most advanced society of the population's ageing. The well-known Ubiquitous (U)-Japan made information and communication ubiquitous tools for disables people using information and communication technology (ICT).

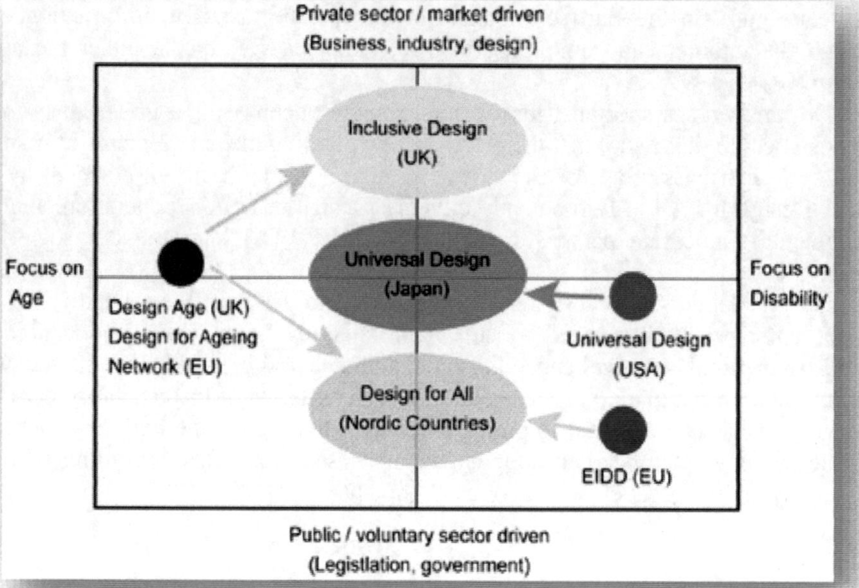

Figure 2. Three origins of influences and ideas on inclusive design

2. Issues related to the lack of Inclusive Design

According to the World Health Organization (WHO) statistics, the rate of disabled people in any country is 10%. The United Nations also published the report accepted this rate in 2021. This rate reaches 15% in the developing countries. Population of Sabah was estimated to be approximately 3.83 million in the year of 2021. According to the statistics issued by Department of Social Welfare Malaysia in January 2021, there were 37,971 persons with disabilities (PWD) based on age category in Sabah, which contribute the demand of public buildings complete with facilities for persons with disabilities (PWD), small children, elders and caretakers. Therefore, the planner in need to review the Inclusive Design in public buildings.

Disabled people are 'persons with physical, mental and intellectual disabilities that inhibit them from fully participating in a normal movement or activities in the community way of life' as stated in The Malaysian Persons with Disabilities Act 2008 (Act 685). Many scholars concluded that the inaccessible environment issues may affect the wellbeing of people with disabilities (PWDs) [5].

Barriers in architecture restrain people with disabilities (PWDs), it causes difficulties while using the facilities provided in the buildings and from getting in and out of the buildings. The inaccessible environment surrounded by public may also cause unsafe, vulnerable stress, low self-esteem, and discomfiture to the people with disabilities (PWDs) [5]. Therefore, it is important to construct public buildings with barrier free design, it will promote and encourage integration as well as a free access through practical plan and secure measurement. Therefore, it can be used by everyone including those who have visual or hearing impairment, parents who bring baby stroller, wheelchair users, and the elderly. With convenient accessibility, people with disabilities (PWDs) may participate in the social and economic mainstreams without feeling being left out or discriminated.

3. Design Criteria of Public Buildings

Public Building are the buildings that reflect the social, cultural, and economic status of the society. These buildings, which are open and used by everyone and are designed with social identity. Throughout decades, people's tendency to come and socialize together, the change of lifestyles has also caused changes in the design of public buildings.

Contemporary societies with an increasing quality of life over time now are demanding more comfortable and reliable buildings. Public buildings accessible to everyone have important considerations on urban area where they are located. The design of public buildings should be suitable for everyone's accessibility in every level of design. Therefore, "approach to building", "Circulation within building", and "reaching services in spaces" must be organized appropriately". Figure 2 illustrates the design scales in architecture according to [13].

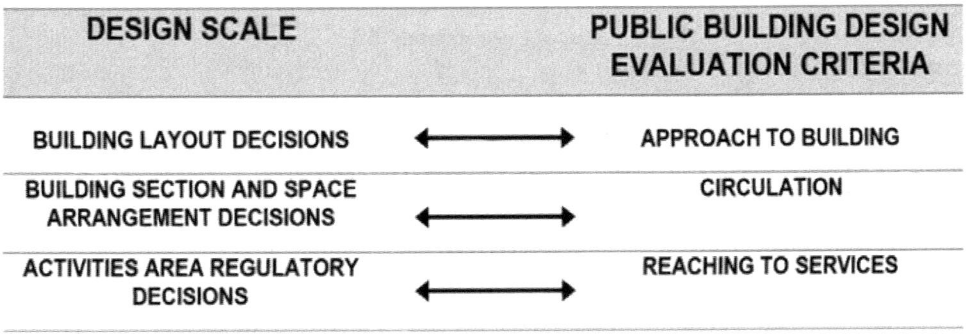

Figure 3. Relationship between design scale and public building design evaluation criteria

Circulation deals with building entrances and exits, including horizontal and vertical circulations within the building. Reaching services examines the spaces that make up the building parts; in other words, it questions each user action area's functional arrangement. Hence, evaluation criteria of public building design are important. Table 1 lists detail evaluation criteria which used as a reference for the evaluator. Next section shall discuss each of the criteria.

Table 1. Public Buildings accessibility evaluation form

Name of Public Building:		
Building Purpose:		
Access	Item	Question
APPROACH TO BUILDING	Transportation Status	• In which ways can access to the building be provided? • Are there any stops or drop-off points for public transportation vehicles located near the building?
	Near Environment	• The roads, sidewalk, ramps, stairs in the immediate vicinity of the building are positioned in appropriate places, It is suitable for everyone?
	Car Park	• Do parking lots and pedestrian crossings meet the standards. It is securely regulated for everyone's use?
CIRCULATION	Building Entrance	• How many entrances are there in the building? • Are the entries detectable and usable by everyone? • Are the doors in the building suitable for everyone? What are the vertical circulation elements in the building?
	Main Entrance Door	• Is the main entrance door perceptible and usable by everyone?
	Horizontal Circulation	• Are corridors and halls suitable for use by everyone for horizontal circulation?
	Vertical Circulation	• Are vertical circulation elements suitable for different user groups?
REACHING TO SERVICES	Information Desk and Other Counters	• Are the counter heights suitable for different user to reach?

Waiting Area	• Do the waiting areas allow different users to wait together under the same conditions?
Wet Spaces	• Do toilet offer options for users with different characteristic) such as tall, short, children, elderly, wheelchair users?)

3.1. Approaching to Public Buildings
All individuals living in the area can easily access all of the city's public services. This is the most natural right of every individual sharing the city. Various users such as individuals with mobility restrictions, children, young people, the elderly, pregnant women, parents using strollers, disabled people use vehicles in the city. All of these users should be able to safely and comfortably use public transport in the city. Appropriate designs should be made for all users in pedestrian roads, vehicle roads and their surroundings, stops and stations, sidewalks, stairs and ramps, parking lots, urban equipment, and other uses in the city.

Some improvements in fulfilling the requirement of persons with disabilities (PWD) has been seen in Malaysia. As reported by Kamal Malhotra (2010) [5], there were great requirement to competently implement universal design as it is a part of Inclusive Design in Malaysia. Hence, more professional planners or researchers are necessary needed to re-evaluate the latest updated standards codes used in the existing public buildings.

3.2. Circulation in Public Buildings
A convenient access should be well defined for everyone to see at the entrance of the buildings. Access to the building entrance from the sidewalks and parking lots in the immediate surrounding should be clearly visible. Any level difference between the immediate surroundings and the building entrance doors should be solved by suitable design for users. It is essential to ensure circulation inside the building, starting with the building entrance doors' size to suit all types of users.

The diversity of individuals who will use the building should be considered. The circulation between floors is known as vertical circulation provided by building elements such as stairs, ramps, and elevators. For each floor to be suitable for everyone's use, care should be taken that there is no level difference within itself. If there is a level difference, this level should be exceeded so that it does not pose an obstacle for any user. It is essential that horizontal circulation elements such as corridors and halls are suitable for everyone's use and designed by considering the number of users.

Staircase when designing the riser heights; it should be at a platform that can be easily used by various types of users such as children, users with mobility difficulties, the elderly, pregnant women. Staircase with handrails should be designed at different levels for different handrails user heights. The elevators door should also be determined by the number and variety of users. Ramps, which are the other essential elements of vertical circulation, their width and slope should be suitable and safe for different users such as wheelchair users and midwives using strollers. Also, all interior door widths, opening arms, and opening directions in horizontal and vertical circulation should be designed so that everyone can use it. When the horizontal and vertical circulation elements in the building are created, taking into account that the user mass of public buildings is everyone, the circulation is suitable for all users.

3.3. Reaching Services in Public Buildings
Access to the building and circulation within the building should be provided, everyone able to access the building services offered according to the function, and thus it is an essential criterion in the design of public buildings and creating spaces that can meet the user's needs. A set of guidance and warning equipment for easy access in buildings should be provided with clear information, that will bring the user to the place he wants to go without any danger or negative situation.

These are arrangements that can be understood with the senses of touch, sight, and hearing to serve individuals with various disabilities. These applications are especially important for visually and hearing-impaired users. For example, considering that visually impaired individuals are among the elevator users, audible warning systems should be installed in elevators. The buttons required for elevator use must be understandable by everyone and must have perceptible surfaces to suit visually impaired people. Also, elevator buttons should be positioned at a height for wheelchair users and other types of users, such as short-stacked users.

Information desks to provide information and guidance to users should be located close to the entrance doors or other desks for different purposes in the building designed for all users. The counters should be designed by considering the employee's types and actions who is the permanent user of the counter and other temporary users. Designing counters with sections of different sizes and features suitable for different users and various user activities will enable the counter right design for everyone's access. The presence of toilet groups within the building that other users can access and meet their needs can also provide access. Appropriate public buildings are designed with everyone's access to the building, horizontal and vertical movement within the building safely and comfortably.

3.4. Code of Practice on Access for People with Disabilities (PWDs)

Approaching to buildings, circulation and reaching to services must be considered to make designs suitable for everyone's use. There are some laws, regulations, standards, guidelines, and measures in all developed countries for achieving this. In the rules and declarations of many nations, all individuals in the society have equal rights. This equality should also manifest itself in spaces open to everyone.

Malaysian Standards and practice code were adopted from the international standards. In this standard, regulation for disabled people refers to MS 1184: 2014 that focus for comfort ability of facilities for disabled people in the country. Malaysia Standards are developed through agreement by committees which comprised of balanced representatives from of planners, users, consumers and related stake holders.

The current guideline for local buildings and facilities to permit access to disabled people has been specified in the Malaysian Standard 1184: 2014 - Code of Practice on Access for People with Disabilities (PWDs) for Public Buildings. However, the implementation acts have yet to be established in order to make it mandatory for building owners to provide the facilities based on the guidelines. Some service providers and developers have been neglected the provisions by making excuses on the extra costs that can be saved. The requirements for barrier free area design are included of open spaces, recreational areas and pedestrian routes. It introduces solutions to the principal problems in the design of an accessible outdoor environment.

According to Uum Al-Qura University (2007) [5] there are five major elements have to be considered under the designing phase of public building (commercial complex) for disabled as shown in Table 2 are:

Table 2. Major Elements of Design for Person with Disabilities (PWDs)

No.	Elements	Considerations
1.	Obstructions	Design a barrier-free path for the safety and independence of disabled people, especially the sightless.
2.	Street furniture	Without obstructing, the free passage of pedestrians along travel routes and the accessible amenities are design for people's convenient.
3.	Curb ramps	Overcome changes in level between the pavement and the road surface and also on the pavement itself.
4.	Pathways	Provide clear, obstruction-free, level and wide pathways for the convenience of all users, especially the sightless and people with mobility problems.
5.	Parking	Provide accessible parking facilities as close as possible to the point of destination.

4. Public Buildings in Kota Kinabalu

Kota Kinabalu is the capital city of Sabah in East Malaysia. Public buildings were important for the city and was design and built to improve quality urban living and environment. It is a coastal city on the west coast of Sabah and surrounded by partly sea views and rainforest. It is known for its wet and dry markets, modern boardwalks, beaches, waterfronts and recreational parks, it is also a gateway to Kinabalu Park the World Heritage Site designated by UNECSO in December 2000 and beacon thousands of local and international tourists yearly. This indicates that public buildings are the major attraction to local and international tourists as well as the resident. Therefore, it is significant to ensure all public buildings are built with Inclusive design and accessible to everyone.

As one of the fastest growing cities in Malaysia, Kota Kinabalu often referred as KK, with numbers of public buildings which taking places within the central city and sub urban district. As it is well known, has seen massive changes to the city skyline over the last 15 years. New buildings including mixed use development, shopping malls, hotels, public halls, public transport station, skybridge, Small office home office (SOHO), convention hall and et cetera has been built with accessibilities as part of Inclusive Design for persons with disabilities (PWDs), small children, elders and caretakers.

The main objectives of this study were to identify several elements of inclusive design in public buildings, to determine the inclusive design implementation of public buildings as well as to define the impact of the persons with disabilities (PWDs), small children, elders and caretakers' requirement in Kota Kinabalu.

5. Types of Public Buildings

There were six types of public buildings in Kota Kinabalu has been identified during this study and often used by urban citizens and often by persons with disabilities (PWDs), small children, elders and caretakers. This section shall name some of the buildings categorized under such types.

5.1. Commercial Buildings

Commercial buildings are buildings where business activities take place for commercial purposes. The building which its floor space at least 50 percent is used for commercial activities, such as retail, the providing of services, or food service (restaurants and cafeteria) and indoor recreational center. For examples Center Point Sabah (CPS), Imago shopping mall, Suria Sabah, KK Times Square, Wisma Merdeka, Welcome Seafood Restaurant, Tong Hing Supermarket, Pasar Kraftangan Kota Kinabalu, Pasar Ikan SAFMA, Bataras and Servay Hypermarket.

5.2. Administrative Buildings

Administration buildings are used as offices or for administrative or related purposes by Malaysian Federal, state government, local government secretariats, a statutory body, a board, a council or an organisation established for public purposes, and includes a courthouse or a police station. For examples Mahkamah Tinggi Kota Kinabalu, Yayasan Sabah, Pejabat Pos Kota Kinabalu, Pusat Pentadbiran Kerajaan Negeri Sabah (PPNS) and Pejabat Polis Karamunsing.

5.3. Health Facilities

Health facilities are places that provide health care services. They include government or private hospitals, wards, family planning centers, confinement centers, paediatric clinics, clinics, outpatient care centers, and specialized care centers, such as birthing centers and psychiatric care centers. For examples Hospital Wanita dan Kanak-Kanak Sabah, Hospital Queen Elizabeth, Klinik Kesihatan Luyang, KPJ Specilist Hospital, Jesselton Medical Center and Glengeagles Kota Kinabalu.

5.4. Community or Social Buildings

Community or social buildings are place to a practices social directed toward the creation or enhancement of community among individuals within a geographical area (neighborhood) or with a common need or interest. For examples Dewan Serbaguna Kota Kinabalu, Dewan Foo Chow, Dewan Serbaguna Damai and Auditorium Yayasan Sabah.

5.5. Educational Buildings

Educational buildings are any building used for study or gaining knowledge such as a university, college, technical institute, school, academy, research laboratory, lecture hall, convent, monastery, public library, public art gallery or museum. For examples Universiti Malaysia Sabah, Muzium Negeri Sabah, Kolej Komuniti Yayasan Sabah, Politeknik Sabah, SMK Teknik and Sabah Art Gallery.

5.6. Religious Buildings

Religious Buildings are place of worship through activities associated with a non-profit organization. The purpose is the instil of religious values, serves primarily persons who share the religious belief of the organization. For examples Masjid Bandaraya, Sacred Heart Cathedral, Gurudwara Shaib Kota Kinabalu and Pu Tuo Si Temple.

6. Overview of 1980s and 2000s public buildings in Kota Kinabalu

This section shall discuss the overview comparison of two public buildings in Kota Kinabalu based on the existing implementation of Inclusive Design, namely Karamunsing Complex (shopping mall with office tower) which has been built and completed in 1985 and Sabah International Convention Centre (SICC) which has been design, construct and completed in 2020.

6.1. Location of the buildings

Karamunsing Complex is located at the eastern part of Kota Kinabalu city facing the hill side view and it is one of the earliest shopping malls mixed with office towers in city together with Wisma Merdeka dan Centre Point Sabah. It comprises of four storey of shopping area, sub-basement shopping area, 10 storey of office towers, exhibition hall and multi-storey of car parking.

Sabah International Convention Centre (SICC) is located at the western part of Kota Kinabalu facing the sea view. The Sabah International Convention Centre (SICC) is the largest waterfront purpose-built in East Malaysia. It comprises five stories multi-functional complex, a convention hall with a compactible partition wall system which can divide hall into several rooms. On another level, three large size exhibit halls encompassing a flexible event space, meeting rooms and private lounges are provided. The top floor is dedicated to individual private meeting rooms, a performance arts hall with seating capacity of 1,250 in two-tier amphitheatre accompanied by VIP rooms and lounges. An outdoor plaza can be used for open concerts and exhibitions are connecting to the main lobby. The cafeteria is located on Level 3 which can be access from the main lobby.

6.2. Approaching to the buildings

Both Karamunsing Complex and Sabah International Convention Centre, users and employees, who are permanent users of the building, prefer to reach the building with their private vehicles for the reason sufficient car parks. There are covered disabled car parking at Level 1(Ground Floor). Visitors prefer public transports such as bus, taxi or e-haling services. There is a designated city bus stop along side of the min road and drop-off at the main entrance of the building. After accessing the building, users can use main entrance at Level 1(Lobby) and Level 3(Parking area) in Karamunsing Complex.

A concrete ramp with 1500mm wide accessible by wheelchair was used to enter the main entrance at Level 1 which is designated for covered parking area for both public buildings and Level 3 for Cafeteria at Sabah International Convention Centre (Figure 4 and 5). The main entrance door of the building was emphasized by Modern design with an automatic sliding glass door. There is a wide corridor heading to each halls and lift lobby area.

There are double-swing doors and Automatic sliding glass doors, aluminium-framed glass doors in some places between the hallways and corridors with width of 4000mm. There are corridors to ensure horizontal circulation within the building. There are visible signages with short texts and symbols on the wall or room doors to indicate the purpose of the use of each room or area.

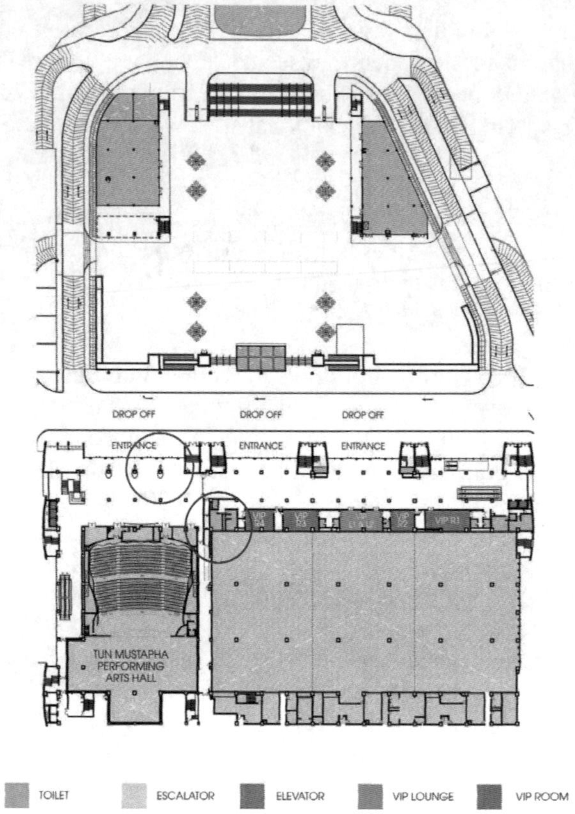

Figure 4. SICC Main Lobby Floor Plan (Level 3)
Red Circle (Counter) Blue Circle (Disabled Toilet)

6.3. Circulation in the buildings

Both Karamunsing Complex and Sabah International Convention Centre provided an Information Counter in different places in the building for information and transaction. The counter, which is approximately 900mm in height, is located at the left side of main entrance and near the lift area. It is located at the entrance (Figure 4) in red colour circle and fixed at an appropriate level so that users can write while standing (Figure 6). On all building floors at Sabah International Convention Centre where the public is served in different departments, waiting area are available. Waiting areas were positioned as an island concept by placing seating furniture in the open large lobby.

6.4. Reaching Services in the buildings

There are many shops, office, halls, function rooms on the floors. Each level provided accessible disabled toilets with signage on the wall at Sabah International Convention Centre (Figure 4) blue colour circle and (Figure 7) photo of the toilet signage. The toilet arranged for the disabled are for one-person uses. There is one sink and wash closet inside complete with disabled sanitary accessories. Grab handles are designed next to the water closet. The mirror is positioned inclined for the individual to see himself.

Figure 5. External Ramp at SICC

Figure 6. Information Counter at SICC

Figure 7. Toilet Facilities Signage at SICC

There is cafeteria in the building located at sub-basement in Karamunsing Complex and the Level 3 at Sabah International Convention Centre, accessible directly from the Main Entrance. Designed with double volume, a reinforced concrete system, brick, solid panel, decorative wall and high wall glass that make up the space. Exhaust Mechanical Fans and air-conditioning are provided at the Karamunsing Food Court for better ventilation (Figure 8).

Figure 8. Karamunsing Complex Food Court (Sub-basement)

7. Conclusions

This paper provided a better understanding of Inclusive Design implementation for public building in Kota Kinabalu. According to the overview investigations made in the study area in each types of public buildings, it was seen that the public transportation, the sidewalk, pedestrian crossings, which are the elements of urban pedestrian circulation, should be improved with an Inclusive Design. Providing convenient access to the building by public transport or private vehicles is vital.

In terms of the building access, it was identified that the furniture used was not suitable for all ages or individual. Therefore, it was suggested to design ideal furniture to everyone's use. It is necessary to interpret the building and surroundings with a contemporary understanding and design space suitable for Inclusive Design. With the Inclusive Design concept in aiming to provide appropriate approach to building, circulation, and access for reaching the services in a building for everyone, public building will emerge for users of all ages especially for persons with disabilities (PWDs), small children, elders and caretakers.

Lastly, all owners of public buildings built before the enactment of the Persons with Disabilities Act 2008 were encouraged to upgrade and to provide basic requirements of inclusive design elements for the convenience of all users.

References

[1] Clarkson PJ and Roger C 2015 History of Inclusive Design in the UK *Applied ergonomics* 46 235-247

[2] Hashim AE, Samikon SA, Ismail F, Kamarudin H, Jalil MNM and Arrif NM 2021 Access and accessibility audit in commercial complex: effectiveness in respect to people with disabilities (PWDs) *Procedia-Social and Behavioral Sciences 50 452-61*

[3] Bashiti A 2016 Accessibility of shopping malls for persons with disabilities (PWDs) in compliance with Malaysian standards *(Master's thesis, Gombak, Selangor: International Silamic University Malaysia)*

[4] Jabatan Kerajaan Tempatan, Kementerian Kesejahteraan Bandar, Perumahan dan Kerajaan Tempatan 2014 Senarai Semak Kemudahan Akses Berdasarkan Rekabentuk Sejagat MS1184:2014

[5] Kadir SA and Jamaludin M 2012 Applicability of Malaysian standards and universal design in public buildings in Putrajaya *Procedia-Social and Behavioral Sciences 36 659-669*

[6] Rahim AA, Zen I, Samad NAA and Rahim CRC 2014 Universal Design and Accessibility: Towards Sustainable Built Environment in Malaysia. *Universal Design 299-306.*

[7] Kementarian Pendidikan Malaysia 2020 Buku Data Pendidikan Tahun 2020

[8] Jabatan Perangkaan Malaysia 2018 Siaran Akhbar Penerbitan Buletin Perangkaan Sosial,

Malaysia 2018
[9] Rahim AA, Zen I, Samad NAA and Rahim CRC 2014 Universal Design and Accessibility: Towards Sustainable Built Environment in Malaysia. *Universal Design, pp.299-306.*
[10] Faizan R. 2018 Inclusive Design Assessment of State Park Buildings in Michigan for the Elderly: An Empirical Study Using Case Studies and Focus Group. Michigan State University.
[11] Salleh SF 2018 Masalah pengajaran guru dalam program pendidikan inklusif di sekolah. *Asian People Journal (APJ)1*(2)243-263.
[12] Libertun de Duren N, Salazar JP, Duryea S, Mastellaro C, Freeman L, Pedraza L, Rodriguez Porcel M, Sandoval D, Aguerre JA, Angius C and Ariza MC 2021 Cities as Spaces for Opportunities for All: Building Public Spaces for People with Disabilities, Children and Elders (No. IDB-MG-859)
[13] Ajobiewe T, Adeleye O and Shaibu V 2020 Access to public buildings by the physically challenged in Ibadan Metropolis Education 2024
[14] Building and Construction Authority 2013 Code on Accessibility in the Environment
[15] IDB 2021 Cities as Spaces for Opportunities for all: Building Public Spaces for People with Disabilities, Children and Elders.
[16] Fariz AR 2014 Pelaksanaan Program Disability Equality Training (DET) dalam Memberikan Pengetahuan dan Kesedaran kepada Masyarakat Mengenai Isu Kekuranganupayaan.
[17] Rahim AA and Samad AA 2010 Accessible built environment for the elderly and disabled in Malaysia: Hotels as case studies. *Journal of Construction in Developing Countries, 15(2) 1-21.*
[18] Tah IHM 2014 Orang Kurang Upaya (OKU) dan Hak untuk Hidup: Satu Analisis Perundangan di Malaysia.
[19] Kementerian Pembangunan Wanita, Keluarga dan Masyarakat 2018 Laporan Statistik Social Welfare Statistics Report
[20] Tahir Z, Thambapillay S, Mohd Yusoff JZ and Abdul Rahman AS 2020 Laws concerning disabled persons: an analysis of legal development in Malaysia. *The Malaysian Journal of Social Administration 14(1)96-114*

Acknowledgments
This research is supported by PAM Sabah Chapter, Kota Kinabalu City Council, Department of Social Welfare, State Education Department and all public buildings in Kota Kinabalu.

IOP Conf. Series: Materials Science and Engineering 1229 (2022) 012007 doi:10.1088/1757-899X/1229/1/012007

Ground track response behaviour induced by train speed using seismic wave method

N N Yahaya[1], A Ibrahim[2,*], J Ahmad[3], A Ahmad[4], M I F Rozli[5] and Z Ramli[6]

[1] Postgraduate Student, School of Civil Engineering, College of Engineering, Universiti Teknologi MARA, 40450 Shah Alam, Selangor.
[3] Senior Lecturer, School of Civil Engineering, College of Engineering, Universiti Teknologi MARA, 40450 Shah Alam, Selangor.
[2,4,5] Senior Lecturer, School of Civil Engineering, College of Engineering, Universiti Teknologi MARA, Cawangan Pulau Pinang, Kampus Permatang Pauh, 13500 Permatang Pauh, Pulau Pinang
[6] Engineer, Keretapi Tanah Melayu Berhad, Pejabat Jurutera Wilayah Utara, 14000 Bukit Mertajam, Pulau Pinang.

E-mail: ceanas@uitm.edu.my

Abstract. Ground track response analysis is an alternative method utilized to investigate the influence of ground-borne vibrations induced by speed train on track. This study intended to get an understanding about the responsive of ground towards vibration induced from moving train. To facilitate this study, non-destructive seismic wave method was performed using new application, Sirius M instrument to identify the peak vertical acceleration generated at various running speed and track locations. The result shows the vertical acceleration data signalized from transition wave generate through passing Electric Train Service (ETS) with maximum 140km.hr^{-1} speed train is higher than commuter with 120km.hr^{-1} speed train which are 1.935m.s^{-2} and 1.051m.s^{-2} respectively. Analysis of vertical acceleration data based on different track locations corresponding to the ETS speed resulting higher peak acceleration at stable track, KM21 compared to settlement susceptible track, KM20.75 which are 6.565m.s^{-2} and 1.935m.s-^{2} respectively. The values obtained from this study indicated ground-borne vibration influenced by speed of train and different type of embankment foundation. This data can be used to assess the influences of train type and speed. Moreover, this on-site ground response measurement is potentially useful as an alternative method to determine the soil stiffness which provide an indication to the possible problematic ground susceptible to the settlement.

1. Introduction

The subgrade soil is important element in the railway track component as it provides stable platform in receiving all the stresses distribute by the above layer. In Malaysia, the subgrade track assessment, particularly in railway engineering has a lack of attention among the researcher as they are more interested/focused on sleeper performance/behaviour on track compared to the other country. Sleeper was considered for irregularities of the train as sleeper crack can be a threat to the overall stiffness. However, the track laid back on problematic soil tends to undergo excessive settlement or uneven level of the track to occur due to pumping effect when subjected to repeated loading as resulting acquire frequent maintenance to keep the quality at the optimum level.

Ground vibration analysis at railway track is the basis measurement in determine soil stiffness as it is crucial to be analyse specifically on train excitation and the resulting of dynamic impact mechanism. The main parameters frequently utilized in modelling prediction of ground response analysis [1]: the axle load and the vehicle speed. The rise in high-speed train become a major concern in environmental vibration along railway lines as the greater track damages possible to occur compared to a normal operational speed train [2]. The presence of vibration movement on track automatically will produce the stress pattern beneath the ground track or known as transition of seismic propagation wave. Upon the seismic wave travels in the ground, the energy or amplitude of the wave will be dissipated along the distance from the loading source and this wave dissipation is the consequence of energy losses in the subsoil which known as attenuation [3]. There is a different between attenuation effect of body wave and surface wave whereby in body wave, the seismic wave dissipated energy losses when travel vertically inside the earth through depth while the surface wave energy dissipates when travel parallel to the ground surface along the distance.

Different speed train will generate diverse wave transition into subgrade layer. The track performance will be under critical speed condition if travelling speed train over the soft subgrade is close to or greater than the Rayleigh wave velocity of the subgrade soil. According to Gao et al. [4], there are two ordinary signs to determine the critical speed condition such as cone-shaped ground wave motion and significant amount of track deflections. Figure 1 below shows the swampy area started to encounter the critical speed effect as the cone-shaped surface wave motion started to grow at 125km/hr of travelling speed train and the shaped become more intense/complex as increasing in speed. However, the site is not under critical speed condition as the rail did not show the excessive deflection thus the deflection contours clearly detected the cone-shaped surface wave motion using field and modeling, even without having any significant increment in rail deflection.

The excessive vibration induced by moving critical speed train which subjected on deposit saturated soil may cause a decrease in effective stress and rearrangement of the soil-particle structure. Both sequences possibly lead to a degradation in undrained stiffness and strength of soils. Previous investigation on soil-structure relationship had discovered by Rozli et al. [5] whereby the higher on operational speed train caused the greater sleeper deflection especially on settlement area of soil due to void existence. Furthermore, critical speed can be classified or characterized as the train speed at which generate higher amplification of the dynamic response of the track structure and subgrade. Critical speed higher than the Rayleigh wave velocity of the subgrade soil plays an essential role in ground track respond.

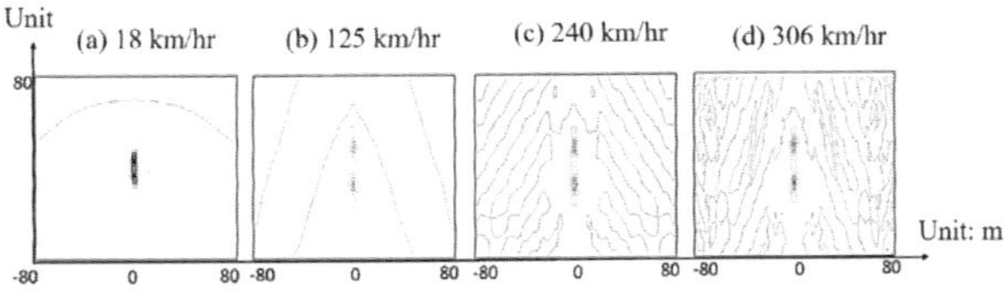

Figure 1. The ground surface wave motion at different speed distribution [4].

Besides, shear modulus can be verified using the seismic wave method. The shear modulus parameter practically used to perform in advanced soil modeling and response in dynamic excitation of the soil-structure interactions. Basically, at low-strain levels, the shear modulus was measured using geophysical techniques that provide the elastic parameter for machine foundation analysis or earthquake engineering. This dynamic property is a resistance to the soil deformation when subjected to shear stress and related to shear strength of soil when subjected to load impact.

Shear modulus of the soil also represents the soil stiffness [6]. The seismic wave method allows to evaluate soil properties of undisturbed condition rather than provide less costly measurement/ economical compared to borehole test. This direct measurement of soil or rock stiffness has the advantage of the material disturbance reduction [7]. Basically, the shear wave velocity of track is crucial to be measured to prevent the uncertainties during analysis the track performance and safety otherwise the measurement of track modulus and shear strength will be estimated. Even though the seismic method frequently used as rapid assessment in monitoring soil profile at subsurface area, Tamrakar et al. [8] and Sussmann et al. [9] found the difficulties in assessing railway track when using spectral analysis of surface wave (SASW) due to presence of ties and rail over the tracks including large variation in ballast and subgrade characteristics. In seismic surface wave testing, geophone and accelerometer are commonly used as receiver wave signal transition.

However, in ballast and subgrade characterization, high frequency accelerometer is essential to receive wide range frequency signal especially for deeper strata. Unlike geophone which strongly discouraged to be used in subsurface determination/analysis due to their mass [8] and lack/ inadequate high frequency response to characterize ballast [10]. Besides, according to Tamrakar et al. [8] mounting effectiveness can influence the quality acquisition data. Major challenge of past study to develop a viable method for fastening or anchoring the mounted accelerometer to detect the generated surface waves and comprehend the effect of wave propagation on track structure [10,11]. As recommended by Tamrakar et al., (2017) to firmly secure the receiver/ accelerometer to the ballast through mounted metal spike for better purpose of SASW utilization.

The main objective of this study is to acquire a comprehensive understanding about the responsive of ground track towards vibration induced from moving speed train. At the same time introduce a novel approach such as triaxial/IEPE sensor that functions similarly to an accelerometer which various used in investigating the ground vibration response toward dynamic impact induce by the train loading. This algorithm can be used as another potential instrument to measure the ground vibration with three-dimensional axis in direct measurement at the same time in identify the modulus elasticity or shear modulus of the subgrade soil in easy way. Meanwhile, the acquisition data will be analyse using the DEWESoft software. Unlike soil behaviour under loading impact, Rozli et al. [5] measured the behaviour of sleepers when they received load waves from a moving train using IEPE sensor with glued directly to the sleeper without influencing the vibration value.

2. Fieldworks and site information

The ground vibration measurements have been performed during passage loading sources of the Electric Train Service (ETS) and Commuter (6 couches) with 140km.hr^{-1} and 120km.hr^{-1} of maximum speeds. The data started to record once the train surpass the track which allocated 12ft distance of the measuring point from the centreline during running test (Figure 2). The tests were carried out at rail track upline at Pinang Tunggal (Tasek Gelugor to Sungai Petani trackline) which recently experiencing the track settlement at KM20.670 (Figure 3) until KM 20.780 whereas the highest settlement achieved at 78mm from track level. The Keretapi Tanah Melayu Berhad (KTMB) maintenance team will normally performed tamping work to realign the track that having soil settlement problem and this maintenance work will sequent use until eventually the track-bed requires full renewal. The tamping process at the same time had increase compaction effect on soil layer during ballast tamping process and the compaction will increase soil density with a consequent reduce air-voids volume but no change in volume of water. The compaction work helps to make soil less susceptible to subsequent volume changes and therefore reduce the tendency to settlement under load or under influence of vibration.

Figure 2. The train loading passage through the track during test.

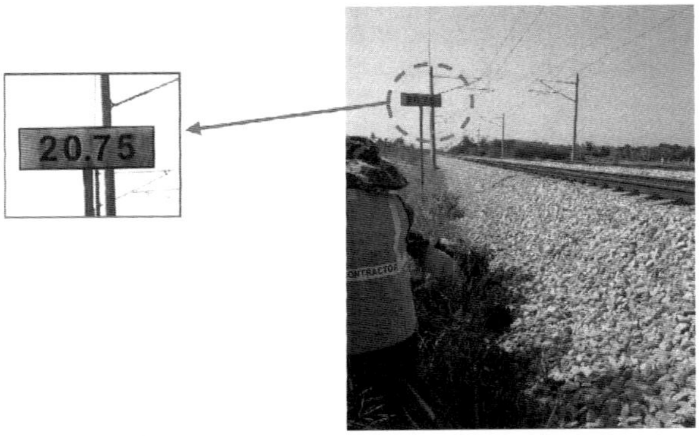

Figure 3. Site selection at KM 20.75 that experienced soil settlement

In late 2007, Yean-chin & Peir-tien, proposed designated ground treatment process to meet the requirement soil strata for double track railway construction project in Peninsular, Malaysia and the geotechnical analysis had been done to investigate the subsoil condition along the existing track. From the site investigation, the subsoil condition for Tasek Gelugor to Pinang Tunggal area were classified as very soft to soft clay type. This result has a correlation with the geological information published by Jabatan Mineral and Geosains Malaysia, about 50% of the track alignment transverse through typical soft alluvium soil along the peninsula region track as shown in Figure 4. According to Lee et al. [12] geotechnical investigation was carried out for the design of embankments of Electrified Double Track over soft clay at Tokai, Kedah. Similar to Pinang Tunggal track area, Tokai site track were underline by the alluvium formation (Figure 4) which relatively homogenous consisting of very soft to soft CLAY (15m thick) overlying dense silty SAND to SAND and approximately unit weight of Tokai clay for top 12m is in the range of $12kN.m^{-3}$ to $14kN.m^{-3}$. Thus, corresponding to the Tokai clay site investigation results, this study applied similar soil unit weight, $14kN.m^{-3}$ to calculate the shear modulus using DEWESoft software and 25kPa of undrained shear strength [13] as data-based references to identify soil condition at both site selection.

IOP Conf. Series: Materials Science and Engineering 1229 (2022) 012007 doi:10.1088/1757-899X/1229/1/012007

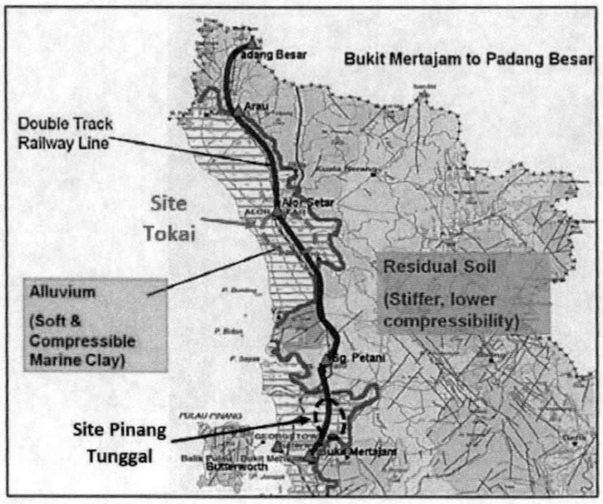

Figure 4. Geological map information on double track alignment (Jabatan Mineral and Geosains Malaysia, 1985) [13]

3. Instrumentation and experimental works

The acceleration data from IEPE sensor was analysed using DEWESoft software. This sensor is able to capture the acceleration data in three directions: vertical, longitudinal and lateral direction. The IEPE sensor is attached on top of metal plate surface using specified 'glue' to improve the accuracy during vertical ground track assessment. The metal plate spike is driven vertically at 100mm depth into the foot ballast using steel hammer to overcome the coupling problem (figure 5 (a)). The horizontal distance of receiver placement at foot ballast is 12ft from centre of trackline and these define location is based on verification by previous studies whereby the maximum acceleration was significantly decreased as the sensor distancing from the rail (Figure 6). As the sensor far distance from track centreline, most of the vibration energy had been dissipated through the moving of wave propagation as considered the damping effects and the internal damping of the subgrade material. This easy installation method introduced/ designed to allow rapid/smooth installation and removal in a short time at multiple sites along the track. The data logger, Sirius M used to transfer the data from the receiver to the portable laptop for further analysed using DEWESoft software (Figure 5(b)).

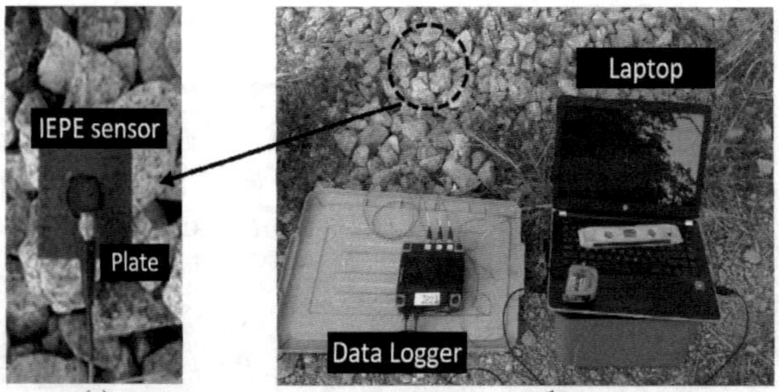

(a) (b)

Figure 5. (a) Close up view of IEPE sensor placement on driven plate and (b) Overview at test area with installation of IEPE connected with Sirius M and portable laptop.

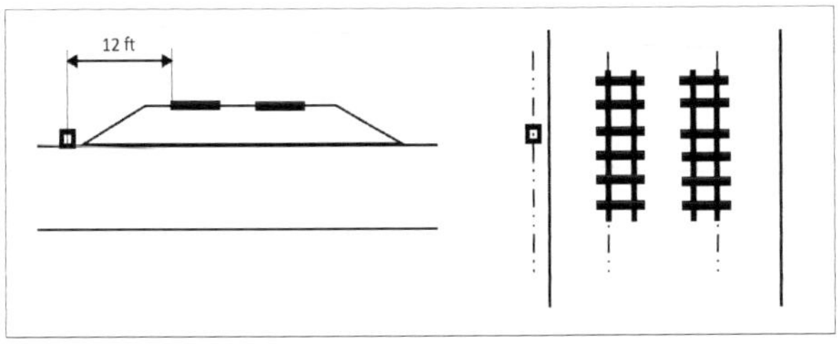

Figure 6. The schematic diagram of installed sensors at track

4. Mathematical integration

According to Zhu et al [14], the vibration signal analysis basically used thc integral method on captured acceleration data to interpret the velocity data as shown in Equation (1).

$$v = \int (a)dt \tag{1}$$

Where

v = velocity (m.s^{-1})

a = acceleration (m.s^{-2})

In order to proceed with the direct calculation of shear modulus (G) of embankment surface [15] mass densities (ρ) obtained from typical unit weights (γ) for clay soil was used to convert the shear wave velocities to shear moduli using Equation (2) below.

$$G = pV_s^2 \tag{2}$$

Where

G = Shear Modulus (kPa)

ρ = soil mass (kg.m^{-3})

Vs^2 = shear wave velocity (m.s^{-2})

According to Stark et al. [10] from point-receiver land/surface-seismic data, the modulus of soil deformation (Young's modulus, E) can be calculate using the following Equation (3) below as directly involve with the assumption of poisson's ratio, v of subgrade layer (Table 3).

$$E = 2G(1+v) \tag{3}$$

Where

E = Young's modulus (MPa)

v = Poisson's ratio

Thus, the modulus elasticity and shear modulus can be directly measured using the DEWESoft software instead conducting laboratory test such as bender element test to measure the shear wave velocity.

5. Result and Discussion

The IEPE sensor is practically capable to capture three different directions (vertical, longitudinal, and lateral) of wave transition from load sources. However, in this study only focus on the vertical vibrational to estimate the soil stiffness at two different site which are KM 20.75 and KM 21.0 under two different type of maximum speed train (140km.hr^{-1} and 120km.hr^{-1}). Type or speed train is one of major factor that influenced the ground track response and through the signalized analysis in Figure 7 had shown the recorded data of vertical acceleration at KM 20.75 for ETS and Commuter. The vertical acceleration generates during passage ETS with travelling speed of 140km.hr^{-1} is higher than commuter which are 1.935m.s^{-2} and 1.051m.s^{-2}. Ground vibration response induced at the lowest value of front head and the end of train as shown in Figure 7. This weak signalize response can be related to the exhibit superposition at the head and tail bogies of the train whereas resulting the strongest impact on ground at the central of the train.

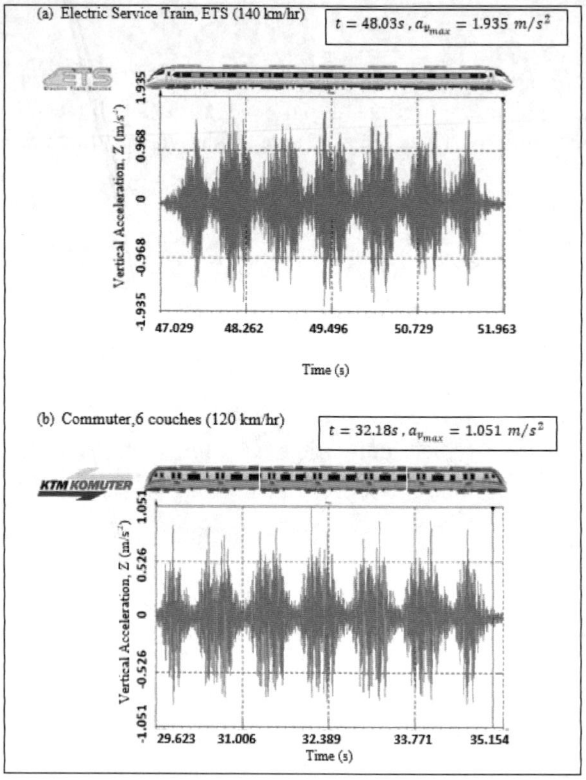

Figure 7. Vertical acceleration signal of ground surface at KM20.75 using DEWESoft software

Table 1. Comparison of acceleration on ground surface under different speed for two types of train at KM 20.75.

Type of Train	Speed (km.hr^{-1})	Acceleration (m.s^{-2})	Percentage Different (%)
Electric Train Services (ETS)	140	1.935	45.68
Commuter (6-couches)	120	1.051	

The percentage different in Table 1 shows an increment about 45.68% of vertical acceleration between different trains with different speed. Repctitive of high ground vibration impact tends to accelerate diffusion effect of subgrade fine into ballast void which allow permanent deformation on track as the vibration induced by high-speed train or heavy axle load impact influence the rearrangement soil skeleton/ particle distribution in subgrade layer. The acceleration analysis was undergone integration process to verify the velocity values of ground response.

Different location resulting different maximum peak acceleration obtained for Electric Train Service (ETS) whereby at KM 21.0 indicate highest value, 6.565m.s^{-2} whereas only 0.488m.s^{-2} of acceleration detect at KM 20.75 (Figure 8). Due to settlement susceptible zone on the area KM 20.75 resulting on the lowest value of wave signalized compared to the stable soil at KM21.0. Theoretically, the settlement area can be related to the soil decreased in volume as increasing in peak acceleration. The mechanism, ballast void infiltration by subgrade particles due to a lack of a filter layer prompted the track settlement. The reposition of soil structure during train loading impact can as it can synchronize with the earthquake phenomenon which ground deposit shaken continuously. However, the particle grained composition for soil layer may vary the settlement and deformation impact as the seriousness of the settlement and deformation effect fluctuate which reflect on soil type. Weak soil tendency high in compressibility thus resulting penetration the ballast particle into soil layer during train loading impact. Other than void/space beneath sleeper or known as sleeper hanging, the soil composition is able to undergo/generate suction during loading unloading phase. Meanwhile during loading impact, the soil particle is in compressed state.

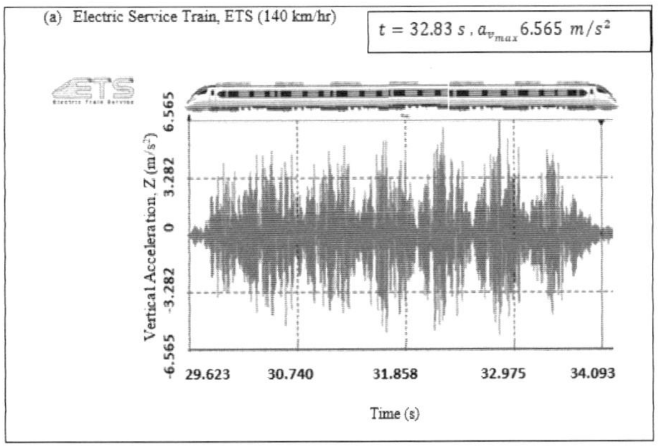

Figure 8. Vertical acceleration signal of ground surface at KM 21.0.

Table 2. Comparison of acceleration value of electric Service Train (ETS) at KM 21.0 and KM 20.75.

Location	Highest Acceleration (m.s^{-2})	Percentage Different (%)
KM 21.0 (Stable)	6.565	70.53
KM 20.75 (Unstable)	1.935	

The acceleration profile for ETS at KM 21.0 shows a high acceleration with 70.53% for percentage different compare with ETS at KM 20.75 as tabulated in Table 2. In previous investigation conducted exactly at similar location at KM 20.75 by Rozli et al. [5] purposely on soil-structure interaction which in turn resulting the highest deflection on pre-stressed concrete sleeper (PCS) recorded at the suspected soil settlement area compared to another site which have stable soil condition. This shows that the soil condition at KM 20.75 still under settlement effect based on time history of acceleration. The problematic soil is crucial to be treated in every railway track as poor subgrade performances can triggered the deflection impact on PCS cracking at worst may lead to track derailment. Therefore, the increase of the train speed partly aggravated the vibrations of track structure which have consequent in accelerate the settlement impact as prompted by larger vibration levels.

In seismic response analysis, the determination of small strain of shear modulus should be consider the influence of cyclic loading, including both the induced effective stress reduction and rearrangement of the soil-particle structure [16]. Soft soil such as clay and silt which have lower shear modulus compared to coarse grained soils as gravel and sand have higher shear modulus. Table 3 tabulate subgrade parameter which influence by the speed train to determine the shear moduli and elasticity of subgrade soil. Following the Equation (2), the velocity data generate via transition of different speed train loading that surpass the track and soil unit weight from previous soil investigation were used to identify the shear modulus of subgrade soil layer. The assumption value, 0.3 of poisson's ratio of subgrade layer had been made in order to calculate the elastic modulus [17]. In Table 2 had shown the estimated subgrade soil properties based on referring to the previous soil investigation data of double track project, the Tokai clay with 14kN/m3 of soil unit weight was used.

Table 3. Subgrade properties at different soil condition

ETS ($140km.hr^{-1}$)	Soil Unit Weight, γ ($kN.m^{-3}$)	Shear Wave Velocity, *Vs* ($m.s^{-1}$)	Shear Modulus, *G* (kPa)	Young Modulus, *E* (MPa)
KM 21.0	14	430	263.87	0.686
KM 20.75	14	340	164.98	0.429

The stability of the track can be justified by analysis the shear wave transition and the elasticity modulus of soil (Es). According to the data provided in the previous study, the undrained shear strength of typical soft alluvium soil was 25kPa which correspond to 0.375MPa of soil elasticity of Tokai clay. Thus, a comparison had been made between the soil elasticity at KM 21.0 and KM 20.75 with the previous Tokai clay data. Data in table 4 shows the stable soil at condition have higher Young Modulus value of 0.686MPa compared to settlement susceptible zone soil track which is 0.429MPa. Stiffness of soil at settlement susceptible zone is close to the Tokai clay indicating the problematic soft clay zone.

Table 4. Subgrade properties at different soil condition

Location ETS ($140km.hr^{-1}$)	Young Modulus, E (MPa)
KM 21.0	0.686
KM 20.75	0.429
Tokai Clay	0.375

Omar et al. [18] had proved the redeveloped settlement formula by Abbiss, [19, 20] had improve the adaptation of calculating settlement in soft clay at Klang using seismic wave method (spectral analysis of surface wave (SASW) and continuous surface wave (CSW) tests) and damping characteristic (plate load test). However, for the present study, only consider half of Omar et al. [18] completed analysis in determine track performance as further plate load test is required to determine the damping characteristic for estimate settlement prediction otherwise, upgraded the Sirius M to advance data logger. In prediction settlement study by Omar et al. [18], the interpreted shear wave velocities for SASW test are 53.7 $m.s^{-1}$ while the corresponding shear wave velocities for CSW test are 57.2$m.s^{-1}$ which shows the soft clay at Klang is lower in soil composition as lower in shear wave velocity compared to soft clay at Pinang Tunggal. As the result, the prediction on settlement at Klang soft clay in 24hr using SASW and CSW tests which are 25.8mm and 22.6mm. If the damping characteristic data were assumed, the prediction on settlement value at Pinang Tunggal will be lower than the predicted settlements at Klang due to higher value of shear wave velocity.

Thus, further ground track analysis using combination field test at different depth were required to investigate the vertical vibrational induced by speed train. It is essential to consider the vibration impact through the wave transition into soil layer as not all the soil properties capable to sustain vibration impact as the soft soil such as clay and silt have lower shear modulus compared to coarse grained soils as gravel and sand which have higher shear modulus. However, the coarse-grained soil need to deal with degradation of the shear strength. Soil plays an important role in sustain the loading especially track which surround by the paddy field area can justified the soil behaviour which weak and need some improvement work to be done. Thus, this information can be used as reference to assess the track safety as considered the traffic and operating parameter such as speed train.

6. Conclusions

Non-destructive test/track assessment for peak ground acceleration caused by speed train (ETS and commuter) was performed at different location (KM 21.0 and KM 20.17) of Pinang Tunggal track, Malaysia. As for measured data basis, characteristic of ground vibration acceleration in time analysis is obtained. Variation of time-domain peak acceleration values and secondary data acquire for estimation previous soil stiffness was for comparison data to analyse the changes in soil stiffness have been investigate. The time-domain peak value of ground vibration acceleration exhibits an approximately linear upward along with the increment of train speed. The farther the observed point from the track centreline is, absolutely influence the reduction of train speed on ground vibration. The analysis of data is limited to vertical acceleration as this angle is parallel to the vertical (maximum) forces induced by train loading impact compared to other directions.

The acceleration peak value recorded shows the influence of vibration activity on railway structure as it can be related to the earthquake engineering. These vibration from speed train influenced the ground track behaviour as the vibration caused the soil skeleton which exposed to the dynamic impact to rearrange into denser state by compaction of the particle. The result proved that the speed of the moving load influenced the dynamic ground response as the higher the speed train accelerate higher amplitude create on the propagation wave induced. The stability of track can be justified by determination of soil stiffness using shear wave velocity generate through peak ground acceleration data. Through the test, current condition of track at KM 21.0 was safe and stable while KM 20.75 need further soil profile analysis to determine the detail sources contribution to the low stiffness of soil.

The overall data shows that the increment of speed train and soil stiffness are major concern as its influence the ground track stability response. Moreover, the seismic wave instrument used was applicable and practical to determine the soil stiffness for geotechnical design and construction based on the agreement result with the previous study for typical problematic soil. The results revealed that the settlement susceptible area is related to the low soil stiffness parameters obtained where can be classified as clay silt soil. This study indicated that is necessary to reinvestigate the soil behaviour or ground response with variance distance from the rail track for preliminary assessment of potential settlement problem.

7. References

[1] Li P, Ling X, Zhang F, Li Y and Zhao Y 2017 Field testing and analysis of embankment vibrations induced by heavy haul trains *J. of Shock and Vibration* pp 1–14

[2] Mezher S B, Connolly D P, Woodward P K, Laghrouche O, Pombo J and Costa P A 2016 Railway critical velocity-Analytical prediction and analysis *Transportation Geotechnics* vol 6 pp 84–96

[3] Dikmen Ü 2005 Modeling of seismic wave attenuation in soil structures using fractional derivative scheme *J. of Balkan Geophysical Society* **8**(4) 175 – 188

[4] Gao Y, Huang H, Ho C L and Hyslip J P 2017 High speed railway track dynamic behavior near critical speed *Soil Dynamics and Earthquake Engineering* vol 101 pp 285–94
[5] Rozli M I F, Safiuddin C M C J M H, Harun M, Ahmad J, Amin N M, Ghani K D A 2019 The behaviour of Prestressed Concrete Sleeper (PCS) sitting on railway track *IOP Conf. Ser. Mater. Sci. Eng.* **615** 012123
[6] Zhang J, Andrus R D and Juang C H 2005 Normalized shear modulus and material damping ratio relationships *J. of Geotechnical and Geoenvironmental Engineering American Society of Civil Engineers (ASCE)* **131**(4) 453–64
[7] Boominathan A 2004 Seismic site characterization for nuclear structures and power plants *Geotechnics and Earthquake Hazards* **87**(10) 1388 – 1397
[8] Tamrakar P, Azari H, Yuan D and Nazarian S 2017 Implementation of spectral analysis of surface waves approach for characterization of railway track substructures *Transportation Geotechnics* vol 12 pp 101–111
[9] Sussmann T R, Thompson H B, Stark T D, Wilk S T and Ho C L 2017 Use of seismic surface wave testing to assess track substructure condition *J. of Construction and Building Materials* vol 155 pp 1250–1255
[10] Stark T D, Dehlin T J, Nazarian S, Azari H, Yuan D and Ho C L 2014 Seismic surface wave testing for track substructure assessment *Joint Rail Conf. American Society of Mechanical Engineers*
[11] Azari H, Yuan D, Nazarian S, Stark TD, Ho C L and Dehlin T J 2014 Surface wave testing for characterization of ballast and foundation layers *Joint Rail Conf. American Society of Mechanical Engineers*
[12] Lee P T, Tan Y C, Lim B L and Nazir R 2016. Some geotechnical properties of Tokai clay *19th Southeast Asian Geotechnical Conf. and 2nd AGSSEA Conf.*
[13] Tan Y C, Lee P T and Koo K S 2016 Construction control chart developed from instrumented trial embankment on soft ground at Tokai of Kedah, Malaysia *Int. Conf. on Transportation Geotechnics vol* 143 pp 548–55
[14] Zhu Y, Jiang W, Kong X, Zheng Z and Hu H 2015 An accurate integral method for vibration signal based on feature information extraction *J. of Shock and Vibration* pp 1–13
[15] Aboye S, Andrus R, Ravichandran N, Bhuiyan A and Harman N 2011 Site factors for estimating peak ground acceleration in Charleston, South Carolina, based on Vs30 *4th IASPEI/IAEE Int. Symposium: Effects of Surface Geology on Seismic Motion* University of California Santa Barbara pp 23 – 26
[16] Zhou Y and Chen Y 2005 Influence of seismic cyclic loading history on small strain shear modulus of saturated sands *J. of Soil Dynamics and Earthquake Engineering* **25**(5) 341–353
[17] Stark T D, Nazarian S, Ho C L and Tutumluer E 2013 Seismic testing for track substructure (ballast and subgrade) assessment for passenger/freight corridors *Joint Rail Conf. American Society of Mechanical Engineers*
[18] Omar M N, Abbiss C P, Taha M R, Nayan K A M 2011 Prediction of long-term settlement on soft clay using shear wave velocity and damping characteristics *J. of Engineering Geology* **123**(4) 259–270
[19] Abbiss C P 1983 Calculation of elasticities and settlements for long periods of time and high strains from seismic measurements *J. of Geotechnique Thomas Telford Ltd.* **33**(4) 397–405
[20] Abbiss C P 1986 The effects of damping on the interpretation of geophysical measurements *J. of Geotechnique Thomas Telford Ltd.* **36**(4) 565–580

Acknowledgments
The authors would like to express the gratitude toward Ministry of Higher Education Malaysia (MOHE) for financial support under FRGS scheme (600-IRMI/FRGS 5/3 (302/2019)). Besides, the authors are indebted to the Research Management Institute (RMI) and Keretapi Tanah Melayu Berhad (KTMB) for their assistance in completing this research.

Effects of RH-WMA Additive on the Design Binder Content (DBC) of Warm Porous Asphalt (WPA)

M. M. Samat [1,*], J. Ahmad [2], N. Omar b [1], N. Mohammad Hanapi [1] and M. M. Mohd Kamal [1]

1 *Faculty of Civil Engineering, Universiti Teknologi MARA, 26400 Caw Pahang, Jengka, Pahang Malaysia*

2 *Faculty of Civil Engineering, Universiti Teknologi MARA, 40450 Shah Alam, Selangor, Malaysia*

*cemo@uitm.edu.my

Abstract. Warm mix asphalt (WMA) in a pavement construction is a technology that is beneficial and contributes towards reduction of the global warming issues. In this research, WMA was used to design porous asphalt pavement with addition of organic additive which is known as RH-WMA. A Dutch modified gradation to suit Malaysian quarry practice was used to prepare the WMA. The RH-WMA additive ranges between 2 to 5 percent was added to base binder of penetration 60/70 to determine the design binder content (DBC) of the RH-WMA mix. The DBC was then determined based on the binder drainage and Cantabrian tests. The binder drainage test was used to set an upper limit while the Cantabrian test for abrasion loss was adopted to determine the lower limit of the DBC. The results for Cantabrian test indicate significant contribution of incorporating RH-WMA additive into WMA to minimize abrasion loss as compared to unmodified mixes. The binder drainage test results showed that the mixes incorporating RH-WMA additive exhibited higher drainage in varying degrees depending on the quantity of RH-WMA. The discussion was made based on target binder content, critical binder content, maximum binder content and the maximum retained binder content of the of the resulted mixes obtained from the binder drainage test.

Keywords: Warm mix asphalt; Porous asphalt; Binder drainage; Target binder content; Cantabrian; Abrasion Loss.

1. Introduction

In the UK, the first synthetic porous asphalt was used to avoid skidding during landings. Its main objective is to allow the water to flow through the air voids and reduce the ponding of water from the road surface. [1]. In Malaysia, PA pavement is relatively new compared to dense asphalt pavement. The first trial PA pavement was constructed along Cheras-Beranang Road and the second trial site is at Jalan Tebrau which is heavily trafficked road. These trial roads were not monitored and the performance is unknown. However, the PA pavement constructed along the Federal Highway performed throughout the design life of service. In Malaysia, the annual rainfall in 2016 reached over 570 mm. [2]. This indicates that the use of PA should improve its skid resistance and minimize ponding. However, ravelling and binder drain down are two common problems which causes inconsistency in the performance of PA [3]. In Malaysia, hot mix asphalt is commonly used to create dense-graded pavement. This process involves heating aggregates at temperatures ranging from 140 to 160 degrees Celsius before they are mixed with bitumen. The high temperatures used during the production and compaction process contribute to greenhouse gas emission. The continuous heating of HMA at high temperature can cause toxic gases to be released into the environment [4]. This issue has been identified as one of the main factors that contribute to global warming. In 1997, the United Nations Framework Convention on Climate Change was discussed to reduce the emission of greenhouse gases as stated in Kyoto Protocol [5]. One way to reduce greenhouse gas emissions is to use warm mix asphalt (WMA). According to Wurst and Putman [6], this material is more energy-efficient and produces less CO_2 emissions.

WMA also consumes less energy and produces less CO2. It's also beneficial for the environment as it allows for longer working hours and better health conditions of the workers. [7].

In Malaysia, a new technology known as WMA is being widely used in the production of asphalt. This process involves reducing the mixing temperatures and improving the binder viscosity. WMA is made by putting an additive into an asphalt mixture to considerably lower the mixing and compaction temperatures. As a result, this strategy can reduce plant emissions, increase field compaction, extend the paving season into colder weather, and lower plant energy use Additives either in organic or chemical form have been used to produce WMA. Silva et al [8] mentioned that the use of wax additives mixed in bitumen helps to reduce the viscosity allowing producing WMA. Chemical additives, such as emulsification agents, surfactants, and antistripping agents, can also be used to reduce temperature, resulting in improved bitumen coating, workability, and compatibility. [9,10,11].

To rationalize the design binder content (DBC) of a porous mix, several trials were conducted. The DBC of the selected gradation is determined based on binder drainage test and Cantabrian test. A value of upper and lower limit of binder content were determined by the binder drainage and Cantabrian test results respectively. The Transport Research Laboratory (TRL) was developed a method based on the binder drainage test in order to determine the upper limit of the DBC as described by Daines [12]. Daines [13] has used this test in a full-scale trial to specify the binder content for a wide range of binder types. In Figure 1, a graphical data shown the typical result of the TRL binder drainage analysis. The maximum value of mixed binder content is the highest value for retained binder. The content of mix binder (M) is assumed to coincide with 0.3% drainage. Therefore, the value for target binder content is equivalent to (M-0.3) %. In principle, the target binder content is the maximum binder content can accommodate the excessive of binder flow during transporting, mixing and laying. While the lower limit of the DBC was determined by using the Cantabrian test based on the method described by Jimenez and Perez [14]. It is important to specify the permitting abrasion loss at ambient temperature for Cantabrian test. At 18°C, 20°C and 25°C, the recommended maximum permitted abrasion loss value for freshly compacted specimens tested are 30%, 25%, and 20% respectively. A graphical relationship from the above data is used as a basis to specify the value of permitting abrasion loss at ambient temperature for Malaysian weather condition.

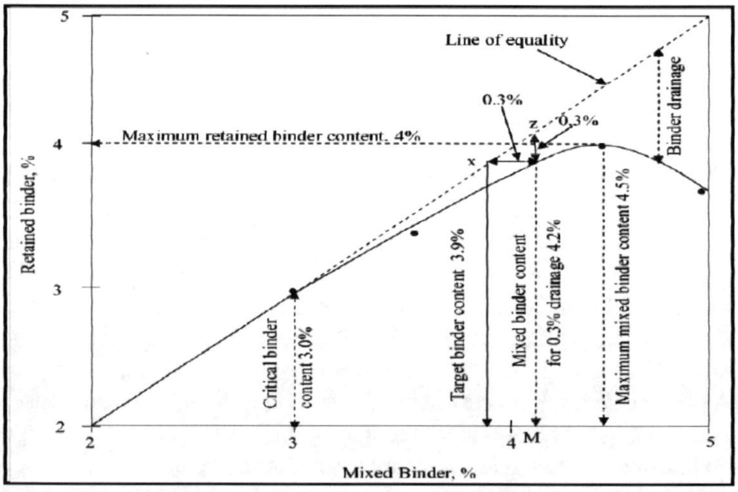

Figure 1. Typical result of the TRL binder drainage analysis (Department of Transportation, 1993)

In terms of mix design, function, and behaviour, there are considerable distinctions between porous asphalt and standard mix. It works as a PA mix at the top of the surfacing layer to speed up surface water removal and prevent water ponding. PA's mix design process is incompatible with standard mix design procedures, which often include the Marshall test. This is because changing the percentage of binder component has no effect on the Marshall Stability value. As a result, specifying upper and lower limit binders using DBC for PA is preferable to specifying the optimum binder content (OPC). The major goal of this laboratory study is to see how different percentages of RH-WMA, a novel wax-based WMA addition, affect the DBC.

2. Methodology

2.1 Materials

In this study, the aggregate material which is granite was supplied by Blacktop quarry in Semenyih, Selangor. A bitumen 60/70 penetration grade incorporating 2%,3%,4% and 5% RH-WMA additive were used to prepare the PA specimen. RH-WMA is a polyethylene wax-based asphalt binder additive produced from cross-linked polyethylene. It was developed by the Research Institute of China Highway Ministry of Transport. In this research, the 0% to 5% RH-WMA additives were used by mass of bitumen.

2.2 Experimental Procedure

The laboratory work is restricted to a study of binder drainage and abrasion loss in a PA mixture known as WPA mix. The DBC's lower limit is determined by an abrasion loss test, whereas the upper limit is required to minimise binder drainage while still promoting high porosity and permeability. The aggregate grading is based on similar gradation developed in Netherlands, which was modified by Aman to suit Malaysian quarry practice [15]. The WPA mix was prepared at varying percentage of RH-WMA additive. The TRL Method was used in investigating the binder drainage characteristic in the laboratory.

2.2.1. Binder Drainage Test Procedure

The TRL procedure described by Daines was used in this laboratory work to conduct the binder drainage test [13]. Table 1 presents the RH-WMA additive contents at 2%, 3%, 4% and 5% mixed with the base binder. The Brookfield Viscometer was used to obtain the mixing temperatures following the ASTM D4402 (2005) [16]. The viscosity of 0.17 ± 0.02 Pa.s and 0.28 ± 0.03 Pa.s was used in the semi-logarithmic relationship between viscosity and temperature for binders with and without RH-WMA content [17].

Table 1. The composition of RH-WMA additive in Base Binder

Binder Type	Binder Content (%)	RH-WMA (%)
Base 60/70	4.0	0
	4.5	0
	5.5	0
	6.0	0
	6.5	0
Base 60/70 + Additive	5.0	2, 3, 4 and 5
	5.5	2, 3, 4 and 5
	6.0	2, 3, 4 and 5
	6.5	2, 3, 4 and 5
	7.0	2, 3, 4 and 5
	7.5	2, 3, 4 and 5

Five metal baskets were created as part of the project. All baskets were made of perforated metal with 3mm diameter holes to allow the specimen's bitumen and filler to drain down throughout the test. Each basket was attached to the steel oven rack via the metal clip with an S-shaped clip at the top and removable wire to hang conveniently on a pre-weighed drainage tray. Preheating the test unit of the mixture in an oven set at the test temperature as specified in Table 2 for at least 4 hours prior to the test was part of the test method. The drainage trays were covered in aluminium foil and placed under the basket on a regular basis. At the mixing and compaction temperatures listed in Table 2, 1100 g of aggregate was mixed with bitumen, integrating with and without RH-WMA additive. The mixer bowl and mixing paddles were scraped thoroughly to verify that all of the mix had been placed into a basket. The operation of placing the mix into the drainage basket was completed swiftly in order to minimise heat loss. The pre-weighed drainage trays coated in aluminium foil were suspended over the metal baskets holding the mix. After 3 hours in the oven, the baskets and trays were removed, and the mixtures in the baskets were discarded. Any evidence of mastic running into the drainage trays was then observed. The mass of drained bitumen and filler also known as a mastic were then weighed after the tray was sufficiently cooled at ambient temperature.
The mass that drained into the drainage tray was a mastic comprising of bitumen and filler. The retained binder (%), R, was calculated using Equation (1) [13].

$$R = \frac{100 X B[1\text{-}D/B\text{-}F]}{1100+B} \qquad (1)$$

where,

D = the weight of mastic drained (g)
B = the initial weight of binder in the mix (g)
F = the initial weight of filler in the mix (g)

2.2.2 Cantabrian Test Procedure

The resistance to disintegration particle was analysed through the Cantabrian test. The test known as abrasion loss was conducted in the Los Angeles rattler, without balls. The Marshall samples were prepared and compacted with 50 blows per face before conducting the abrasion loss test. In the Cantabrian test, a Marshall sample was subjected to 300 drum rotations in the Los Angeles drum at an ambient temperature of 29°C. Abrasion loss is expressed in terms of the percentage of weight loss relative to the original weight as illustrated in Eq. (2). The test method was adopted according to the procedure described by Jimenez and Perez (1990). Samples were tested at binder contents ranging from 3.5% to 5.0% in 0.5% increments and varying percentage of RH-WMA additive starting from 0%, 2%, 3%, 4% and 5%.

$$P = \left[\frac{P_1 - P_2}{P_1}\right] X\ 100 \qquad (2)$$

where

P = Abrasion Loss (%)
P_1 = Initial weight (g)
P_2 = Final weight (g)

3. Results and Discussion

3.1 Effect of RH-WMA additive on bitumen viscosity

Figure 2 presents the temperature-viscosity graph of the base binder containing different percentage of RH-WMA additive using a Brookfield Viscometer conducted according to ASTM D4402 [16]. The mixing and compaction temperatures based on a viscosity of 0.17 ± 0.02 Pa.s and 0.28 ± 0.03 Pa.s was plotted on semi-logarithmic scale [17].

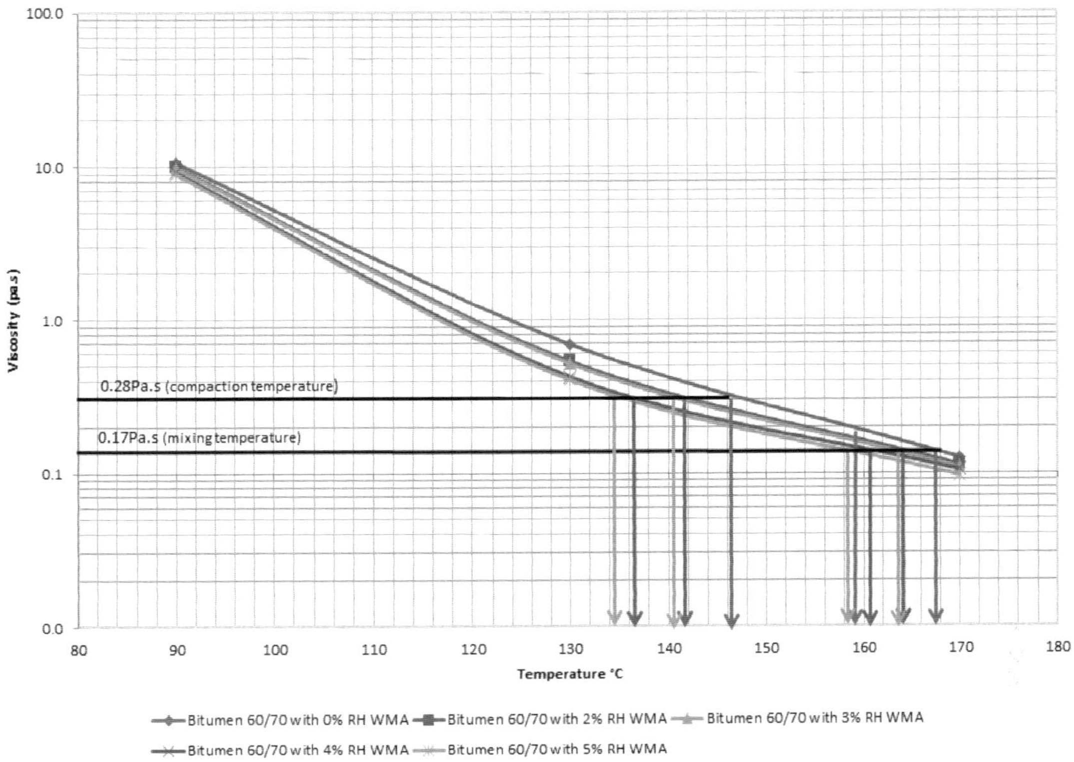

Figure 2. Viscosity-Temperature relationship for binder containing RH-WMA additive

Table 2 summarizes the mixing and binder drainage test temperatures obtained. Based on the viscosity-temperature above, the RH-WMA mixes show much higher viscosity over the range of test temperatures compared to those of base binder PEN 60/70. The mixing and compaction temperature for base binder without RH-WMA is 167.3°C and 159°C respectively. With respect to the different percentage of RH-WMA, results obtained for mixing and compaction temperature is summarized in Table 2.

Table 2. Mixing and Compaction Temperature of binder with and without RH-WMA additive

RH-WMA additive	0%	2%	3%	4%	5%
Mixing temp (°C)	159 – 167.3	154.2 - 164	153.2 – 163.7	148.9 – 160.5	146.3 – 158.3
Compaction temp(°C)	146.8 - 153	141.9 - 148	140.4 - 147	136.4 – 141.9	134.7 – 141.3

IOP Conf. Series: Materials Science and Engineering 1229 (2022) 012008 doi:10.1088/1757-899X/1229/1/012008

3.2 Binder Drainage Test

The binder drainage conditioning or oven temperature is higher than the mixed temperature to ensure that drainage occurs in the oven rather than in the mixer during the mixing process. Table 3 shows the mixing and drainage test temperature. Relationship between the retained binder content versus mixed binder of the WPA mix with and without RH-WMA additive is presented in Figures 3 to 7. If no drainage takes place, then the graph obtained will be a straight-line sloping at 45° degree, which is the line of equality. Figures 3 to 7 show the effect of RH-WMA additive percentages on the extent of binder drainage.

Table 3. Mixing and Binder Drainage Test Temperatures

Binder Type	Mixing temp. (°C)	Binder Drainage Test temp. (°C)
Base 60/70	155	165
Base 60/70 + 2% RH-WMA	155	165
Base 60/70 + 3% RH-WMA	155	165
Base 60/70 + 4% RH-WMA	150	160
Base 60/70 + 5% RH-WMA	145	155

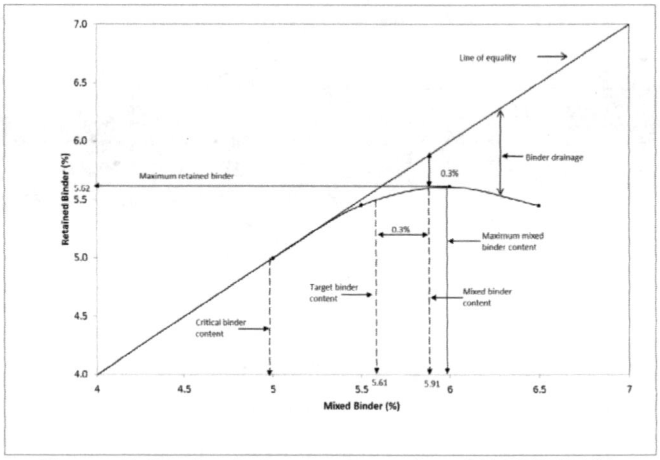

Figure 3. Binder drainage test results - 0% RH-WMA

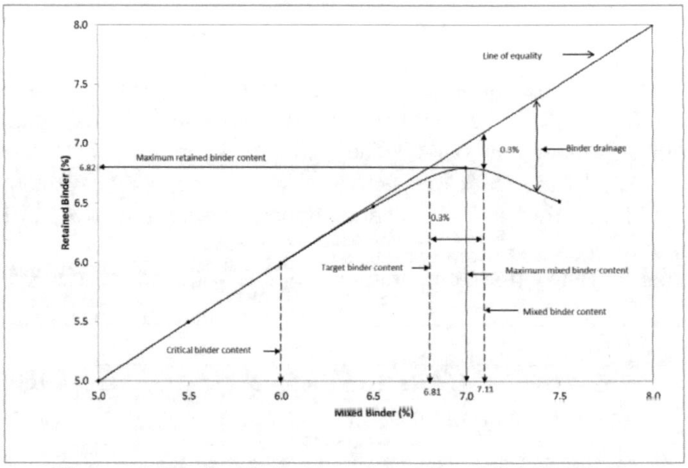

Figure 4. Binder drainage test results - 2% RH-WMA

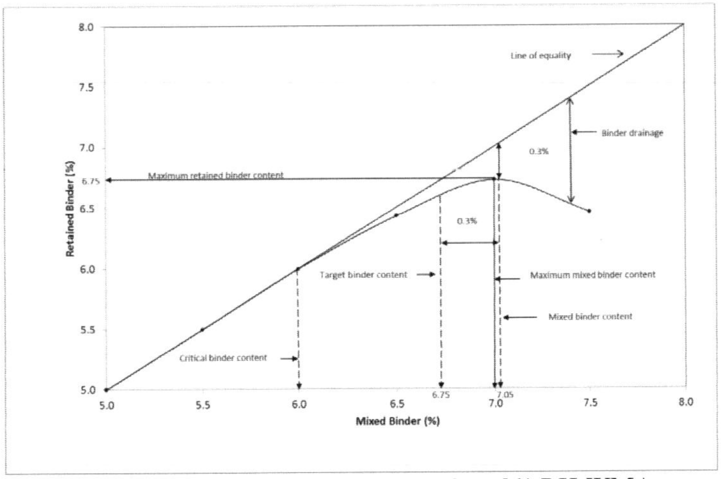

Figure 5. Binder drainage test results - 3% RH-WMA

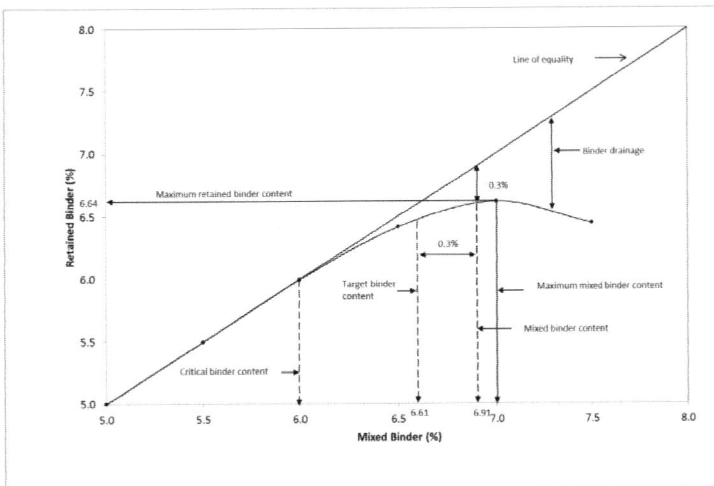

Figure 6. Binder drainage test results - 4% RH-WMA

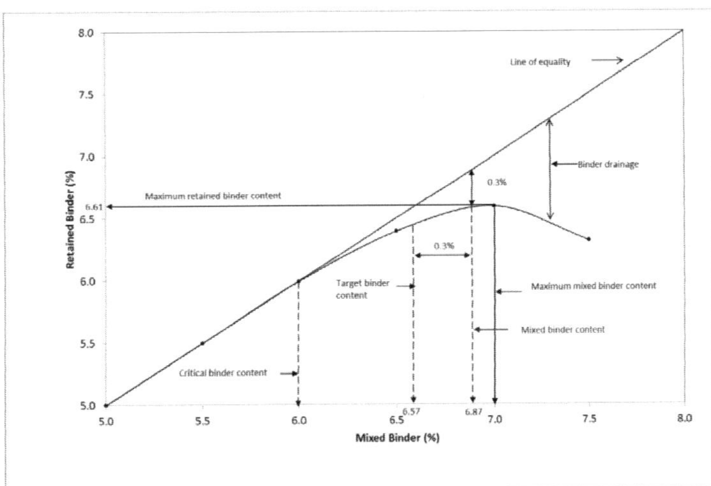

Figure 7. Binder drainage test results - 5% RH-WMA

IOP Conf. Series: Materials Science and Engineering 1229 (2022) 012008 doi:10.1088/1757-899X/1229/1/012008

The binder drainage test results are tabulated in Table 4. The critical binder content is approximately 6.0% and is almost similar for all mix except mix without RH additive. Hence, the onset of drainage is insensitive to modifier content. The maximum mixed binder content indicates the maximum binder content that can be added beyond which binder drainage will take place quite considerably. Without a RH additive, the maximum mixed binder content is found to be 6.0% but increases by 1.0% when the varying percentages of RH-WMA additives were added. The maximum retained binder content decreases as the additive content increases. Without RH-WMA additive, the value of maximum retained binder content is 5.62%. However, the addition of RH-WMA additive from 2% to 5% showed that the retained binder content decreases from 6.82% to 6.61% respectively. This is because the RH-WMA additives reduce the viscosity of asphalt binder. In addition, the binder drain down drastically occurred. Therefore, the increase of RH-WMA additive used can decrease the maximum retained binder content at elevated temperature. This verifies the Putman and Lyons that the binder does migrate vertically through the specimen from top to bottom due to gravity at high service temperatures [18]. The mixed binder content corresponding to 0.3% drainage for all mix occurred below the maximum mixed binder content. Exceptions are for mix incorporating 2.0% and 3.0% RH-WMA additives where the point is located on the peak of the binder drainage curve.

Table 4. Binder Drainage Test Results – TRL Method

Binders	Max. Retained Binder Content (%)	Critical Binder Content (%)	Target Binder Content (%)	Max. Mixed Binder Content (%)	Mixed Binder Content (%)
Base 60/70 + 0% RH-WMA	5.62	5.0	5.61	6.0	5.91
Base 60/70 + 2% RH-WMA	6.82	6.0	6.81	7.0	7.11
Base 60/70 + 3% RH-WMA	6.75	6.0	6.75	7.0	7.05
Base 60/70 + 4% RH-WMA	6.64	6.0	6.61	7.0	6.91
Base 60/70 + 5% RH-WMA	6.61	6.0	6.57	7.0	6.87

The target binder content was set at less than 0.3 percent mixed binder content for design objectives. The target binder percentage for conventional bitumen is 5.61 %. However, as shown in Table 4, adding additives has an influence on the target binder content, whereas the target binder content decreases with RH-WMA content. The target binder content, maximum maintained binder content, and critical binder content % all reduced as the RH-WMA additive in binder content increased. The addition of RH-WMA additives to PA has a considerable impact on the binder content determination. The need of binder material can be lowered by increasing RH-WMA additives in the mix.

3.3 Cantabrian Test

It is important to specify the permitting abrasion loss at 29°C ambient temperature for Cantabrian test. The recommended maximum permitted abrasion loss value for freshly compacted specimens tested at the temperature of 18°C. 20°C and 25°C are respectively 30%, 25% and 20% [19]. The permitting abrasion loss value at 29°C ambient temperature was specified by extrapolating this curve and equals 16.8%. A graphical relationship is shown in Figure 8. This value is the permitting abrasion loss adopted to determine the lower limit of the DBC. Figure 8 shows a graphical relationship between abrasion loss and binder content of WPA mix. The relationship between the abrasion loss and binder content for all mixes prepared with and without RH-WMA additive are shown in Figure 9. In general, all WMA mixes are more resistant to abrasion loss as the binder content increases. For a given binder content, the mix incorporating RH-WMA has higher resistance to abrasion loss compared to WPA without RH-WMA additive.

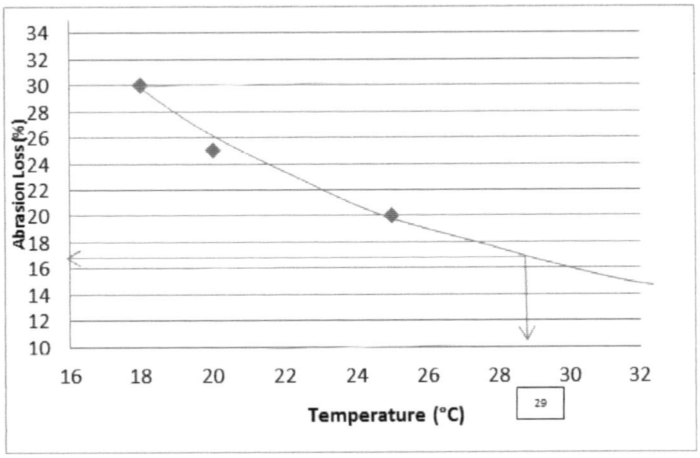

Figure 8. The Permitting Abrasion Loss at Ambient Temperature

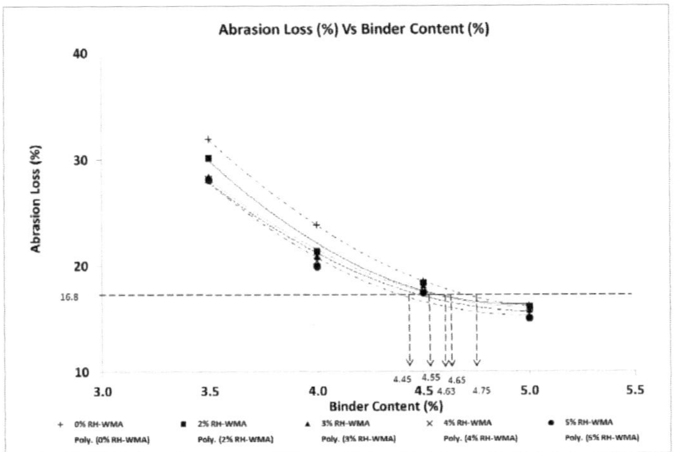

Figure 9. Abrasion loss vs Binder content

Increasing RH-WMA additives content from 2% to 5% has the effect towards the binder content requirement from 4.65% to 4.45%. These binder contents represent the lower limit of the design binder content. Table 5 shows the upper, lower limits from Cantabrian, the binder drainage test and also the design binder content for all the mixes. It shows that increase in RH-WMA additives will reduce the binder content in the mixes. It also indicates that the mean design binder content decreased from 5.73% to 5.51% at 2% RH-WMA and 5% RH-WMA respectively.

The abrasion resistance of the WPA mix in this research was evaluated and the results are summarized in Figure 10. It was found that the amount of binder content exhibited significant influence to abrasion loss. The results showed that 3.5% of bitumen content is not recommended in this study because the abrasion loss of the resulted mixes almost failed. Therefore, 4.0% bitumen is recommended to be used for the lower limit of the binder content. Higher binder content would reduce severe abrasion loss of WPA mix. The increment of RH-WMA additives in WPA mix resulted the abrasion loss meets the maximum specification requirement which is not more than 30% except for the mix without RH-WMA additive and mix with 2% RH-WMA additive.

Table 5. Lower, upper limits and mean design binder content of various RH-WMA

Types of Binders	Lower Limit	Upper Limit	Mean DBC
Base 60/70 + 0% RH-WMA	4.75	5.61	5.18
Base 60/70 + 2% RH-WMA	4.65	6.81	5.73
Base 60/70 + 3% RH-WMA	4.63	6.75	5.69
Base 60/70 + 4% RH-WMA	4.55	6.61	5.68
Base 60/70 + 5% RH-WMA	4.45	6.57	5.51

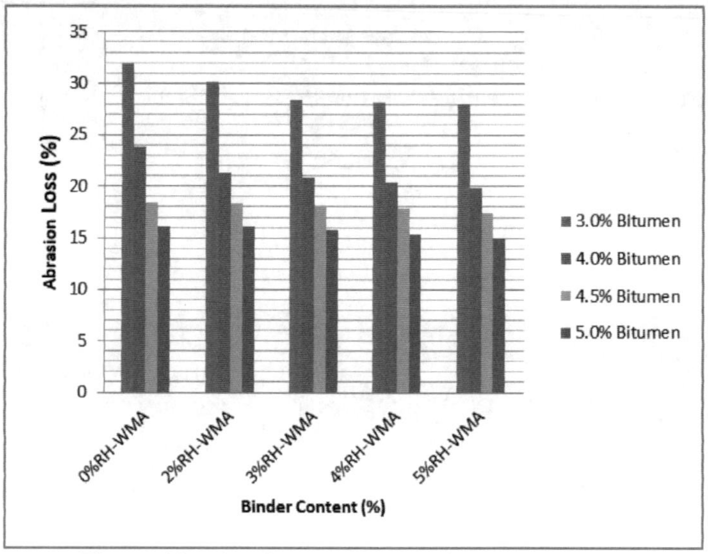

Figure 10. Abrasion loss results of the mix designs

4. Conclusions

To meet the standards, several essential factors must be met while choosing DBC for PA. The binder content should be kept as low as possible to avoid binder run-off and to produce more porous, high-permeability mixes. Furthermore, the binder level should be high enough to prevent oxidation. The binder drainage test determines the maximum amount of binder that a porous mix may hold without causing binder drainage. As a result, it can be used to specify the DBC's upper limit. The drainage properties of WPA mixes made with and without the RH-WMA component varies dramatically. Several conclusions can be drawn from the experimental work, as follows: The required binder content was reduced when RH was added to the WMA mixtures. In addition, increasing RH-WMA content reduced the maximum retained binder content and reduced the target binder content, but the maximum mixed binder content remained constant across all mixes. The Cantabrian test determines the minimum binder content required to ensure resistance to particle loss due to traffic loads. According to the Cantabrian test results, variations in bitumen content are highly sensitive to disintegration resistance, especially in the lower range of bitumen concentration. The inclusion of RH-WMA additives in larger proportions improved abrasion resistance and, as a result, mix durability. This suggests that adding just 2% RH WMA additives to the WPA mix can reduce abrasion loss by up to 5.7 percent when compared to unmodified WMA.

5. References

[1] Hamzah M. O., Hasan M. R. M., and Van Bochove G. Rational control parameters for determination of design binder content of porous asphalt. 5th Eurasphalt and Eurobitume Congress Istanbul; 2012.

[2] Malaysian Meteorological Department Retrieved from Official Portal Malaysian Meteorological Department: Ministry of Science Technology and Innovation (MOSTI); 2021. Available from: http://www.met.gov.my

[3] Kandhal, P. Design construction and maintenance of open-graded asphalt friction courses, Information Series 115. National Asphalt Pavement Association, Lanham, M D; 2002.

[4] Caro, S., Beltran, D. P., Alvarez, A. E., and Estakhri, C., Analysis of moisture damage susceptibility of warm mix asphalt (WMA) mixtures based on dynamic mechanical analyser (DMA) testing and fracture mechanics model. *Construction Build Material.* 2012;**35**, pp 60-67.

[5] Lloyd, B., and Subbarao, S. Development challenges under the clean development mechanism (CDM) – can renewable energy initiative be put in place before peak oil?. *Journal of Energy Policy* 2009;**37**, pp 237-245.

[6] Wurst, J. E., and Putman, B. J. Laboratory Evaluation of Warm-Mix Open Graded Friction Course Mixtures. *J. Mater. Civ Eng*., 2013;**10**.1061/(ASCE)MT.1943-5533.0000611, pp 403-410.

[7] Vaitkus, A., Cygas, D., Laurinavicius, A., and Perveneckas, Z. Analysis and evaluation of possibilities for the use of warm mix asphalt in Lithuania. Baltic *J. Road Bridge Eng*., 2009;**4(2),**pp 80-86.

[8] Silva H.M., Olieveira J.R., Peralta J., Zoorob S.E. Optimisation of warm mix asphalts using different blends of binders and synthetic paraffin wax contents. *Construction Build Material* 2010; 24(9), pp 21-31.

[9] Croteau J.M., Tessier B. Warm mix asphalt paving technologies: a road builder's perspective. In: Annual Conf., of the transportation association of Canada; 2008.

[10] Antonio G.L.J, Lionel G., Gilles B. Warm mix asphalts with low dosage chemical additives. In: 88th ed. Annual meeting of transportation research board, Washington DC; 2009.

[11] Xiao F.P., Punith V.S., Amirkhanian S.N. Effects of non-foaming WMA additives on asphalt binders at high performance temperatures. Fuel 94:144-55; 2012.

[12] Daines M.E. Pervious macadam: Trials on trunk road A38 Burton bypass, 1984. TRRL Research Report 57; 1986.

[13] Daines M.E. Trials of porous asphalt and rolled asphalt on the A38 at Burton bypass. TRRL Research Report 323; 1992.

[14] Jimenez, F. E. P., and Perez, M. A. C., Analysis and evolution of the performance of porous asphalt. The Spanish Experience, Surface Characteristic of Roadways, Int., Research and Technology, ASTM STP 1031, W.E. Meyer and J. Reichert, Eds., Philadelphia 1990; pp 512-527.

[15] Aman, M.O. Water sensitivity of warm porous asphalt incorporating Sasobit®, PhD Thesis, Universiti Sains Malaysia; 2014.

[16] ASTM D4402: Standard Test Method for Viscosity Determination of Asphalt at Elevated Temperatures using a rational Viscometer, American Society for testing and Materials, West Conshohocken, PA 19428–2954, United States: 2005.

[17] Asphalt Institute. The Asphalt Handbook, MS-4., 7th ed. Kentucky, USA, ISBN: 978-1-934154-27-4; 2007.

[18] Putman, B. J., and Lyons, K. R., Laboratory Evaluation of Long-Term Draindown of Porous Asphalt Mixtures. *J. Mater. Civ. Eng.*, 2015;**10**.1061/(ASCE)MT.1943-5533.0001260, pp 91-97.

[19] Colonna, P. Design and performance of porous asphalt in the Pulgia region of southern Italy. Eurasphalt & Eurobitumen Congress; 1996. pp 1-12.

Acknowledgments

The authors wish to acknowledge Fundamental Research Grant Scheme from Kementerian Pendidikan Tinggi Malaysia and Universiti Teknologi MARA, Shah Alam and Universiti Sains Malaysia, Transkerian Pulau Pinang for their financial and technical support of this research.

Physical properties of palm oil boiler ash modified bitumen with Rediset

N Mohd Akhir[1*], B Idiajir[1,2], L Gungat[3], S Harun[4], M N Borhan[1] and N I Md Yusoff[1]

[1] Department of Civil Engineering, Faculty of Engineering and Built Environment, Universiti Kebangsaan Malaysia, Selangor 43600, Malaysia.
[2]Directorate General of Highways, Ministry of Public Works and Housing, Indonesia
[3]Department of Civil Engineering, Faculty of Engineering, Universiti Malaysia Sabah, Sabah 88400, Malaysia
[4]Department of Chemical Engineering, Universiti Kebangsaan Malaysia, Selangor 43600, Malaysia.

*Corresponding author: norliza82@gmail.com

Abstract. This study examines the physical properties of a 60/70 penetration grade bitumen modified using warm mix asphalt (WMA) additive, Rediset and palm oil boiler ash (POBA) as a modifier. Modified binders were prepared by adding 2% of Rediset with different POBA contents (3, 5, 7, and 9%) through wet mixing process. Physical properties of modified bituminous binder were obtained from penetration, softening point, rotational viscosity and ductility tests. The addition of 7% POBA in WMA binder has the best characteristics in term of its physical properties. The penetration values, softening point temperatures and ductility characteristics show inconsistent patterns. However, the rotational viscosity of each combination had decreased with the increment of temperature even though the trend was not significantly constant. Penetration index (PI) showed an increment with 3 and 5% and decrement pattern on 7 and 9% as well as penetration viscosity number (PVN) showing an inconsistent decrement with the addition of POBA. From the results, it can be concluded that using POBA in WMA binder for pavement construction is a viable option and can be further investigated.

Keywords: Bitumen; Rediset; Palm oil boiler ash; Physical properties.

1. Introduction

In current situations, the increased yearly traffic volume has caused so much problem on the overloading, over dimension, accidents and environmental issues that can cause further damage to the pavement, such as increased in the cracking, rutting and other road deteriorations. These issues demand engineers need to finds new inventions regarding improvement in the characteristics related properties of asphalt mixtures. The new innovations in pavement industries like noise reducing asphalt, water-saving pavement, eco-friendly ingredients, perpetual pavements, and computer modelling are driving the future of road technology construction [1]. In order to bear with the advancement in pavement, increased traffic volumes and sustainable technologies, asphalt binders are being modified with various types of modifiers and additive to improve their performance characteristics. In recent decades, there has been a rise in interest in employing renewable resources derived materials (RRDM) to modify asphalt binders [2]. In this study, the RRDM from palm oil industries, for the modification of asphalt binders in warm mix asphalt was investigated. The plant derived materials can be utilized to modify asphalt binders as renewable resources derived compounds [3].

Fresh fruit bunches (FFB) derived from palm oil tree producing palm kernel oil, crude palm oil, and another biomass [4]. Through palm oil milling activities, these FFB generate several wastes includes liquid and solid waste. Solid wastes created by palm oil mill operations that produce crude palm oil (CPO) and palm kernel oil (PKO) are harmful if not properly regulated and monitored. This can result in soil contamination, air and water pollution, and the release of greenhouse gases. Some of these solid waste materials have the ability to be transformed into high valuable product, which might help the palm oil industry grow. Palm oil boiler ash (POBA) is a solid waste generated by palm oil mill industries and it is an environmental issue. After CPO extraction, a massive amount of solid biomass waste is produced, primarily consisting of empty fruit bunches (EFB), palm pressed fibre (PPF), and palm kernel shell (PKS). Approximately 1.77 tons of solid biomass waste is created for every ton of CPO production [5]. Based on average palm oil mill's capacity (10-60 tons FFB/h), an estimated 6-12 tons of solid biomass waste or 65-394 kg ash would be produced on every hour [6].

It is an enormous number of solid wastes; if it is not correctly handled it will pose a severe threat to the environment. Rusbintardjo [7] and Babalghaith [8] studied that POBA is a solid waste that contain a high amount of silica (SiO_2), which has the potential to pollute the ground. As natural radioactive radiation can generate harmful health effects such as cancer and genetic abnormalities that damage future generations, POBA also can be categorized as hazard waste due to its carcinogenic material for human where the level of natural radioactivity potassium 40K higher compared to the standard average range [9]. Therefore, it urges a new invention to turning this ash into a useful final product in order to reduce solid waste from the palm oil industries, moreover lowering the negative environment impact. Previous studies have used POBA as a geopolymer composite material [10], as a heavy metal absorbent in wastewater treatment [11], as an absorbent of water-soluble dispersed dyes [12], as a source of nutrients for fertilizer production [13] and as a substitute for coarse aggregates in high-strength lightweight concrete [14]. POBA was capable of removing 98% of carotenoids and producing refined oil colours that were comparable with conventional bleaching earths [15], POBA geopolymer brick increased in strength over time, with the maximum strength obtained 16.1 MPa and classified as medium weight for non-loading brick type [10], POBA's high organic carbon content improved the effectiveness of nutrient acquisition thus producing quality crop [16].

In road construction industry, POBA initially been paved as road especially at palm oil plantations and mills. Shafii [17] has investigated the potential POBA as a filler in hot mix asphalt mixtures (HMA). However, the author concluded that the POBA unsuits to replace filler in HMA since it has reduced the strength properties of the modified samples. Therefore, in-depth research and study on POBA application in pavement needs to be enhanced in order to reduce the massive amount of biomass. Referring to Alamawi [18], bio-based alternative binders can be used in three ways to reduce crude petroleum sources; as a bitumen modifier (if bitumen replacement is less than 10%), as a bitumen extender (if bitumen replacement is 10-75%) or as an alternative binder (75-100% of bitumen replacement). Bio-oils, additives and waste materials potentially can be used as the alternative for bitumen in the binder mixture [19].

The objective of this study is to investigate the physical properties of a 60/70 penetration grade asphalt binder modified with liquid Rediset as additive while various weight percentage of POBA as modifier (3, 5, 7 and 9%) and to evaluate the optimum percentage of POBA incorporated in the modified asphalt binder. The Rediset is known to improve characteristics performance especially tensile strength [20], anti-rutting properties, resistance to moisture damage [21], stability [22], and fatigue life-span [23]. Rediset was added at 2% of total weight of the asphalt binder. The surfactant effect of Rediset and POBA in bitumen modification were examined in order to improve the internal bonding as well as performance properties of the modified bitumen specifically when produced at the lower temperatures. The WMA binder without POBA (unmodified) is selected as a control samples. The physical test selected in this research included the penetration, softening point, ductility and viscosity tests.

2. Experimental design

2.1. Materials

In this research, palm oil boiler ash would be as the modifier and warm mix additive as the addition materials to the asphalt binder. There are three materials that used in this research, which are asphalt binder, palm oil boiler ash (modifier), and warm mix additive. A 60/70 penetration grade asphalt binder is utilized in the project (herein after referred as Asphalt 60/70). Asphalt 60/70 is choosing because it is mostly commonly used in Indonesia and Malaysia. The source of materials was supplied by Sunway Quarry Industries Sdn. Bhd. in Semenyih, Malaysia. Asphalt binders were stored in closed containers to avoid the effects of oxidation on ageing and held at room temperature and in safe place. The material POBA obtained from Palm Oil Mill in Johor, Malaysia where it consists of large particles which included unburned kernels, nutshells, and fibers as shown in Figure 1(left). The POBA samples contain about 46-47% of moisture, and oven-dried prior to grinding at a temperature of 70°C for 20 hours according to ASTM C70-20 to eliminate the excess water content. The coarse particles of POBA were further ground into finer sizes, followed by sieving process using a 100-mesh sieve size to separate impurities such as sands, stones and macro particles. The powder-sieved POBA (as presented in Figure 1 (center)) was then stored in dry place to avoid contact with moisture. Figure 1(right) shows the surface morphology of POBA. The formation showed a good distribution a pore structure of silica sand (electron image 1) with 1000 times magnifications.

The sieving process had altered the characteristics of POBA. The elemental of composition of POBA is shown in Table 1. In past two decades' warm mix additives have been created to allow asphalt mixtures to be manufactured at lower temperature. Additive used in this study is liquid chemical additive, namely Rediset LQ-1200 (herein after referred as Rediset) would be added into asphalt modified binder. It is a warm mix additive that offers an easy-to-use liquid product that simplifies management and measurement at the asphalt mixing plant. It offers good workability and compaction at lower temperatures. Furthermore, it provides active adhesion which allows coating of aggregates that are not completely dry, prevents stripping and increases resistance to moisture. The physical properties of Rediset are presented in Table 2.

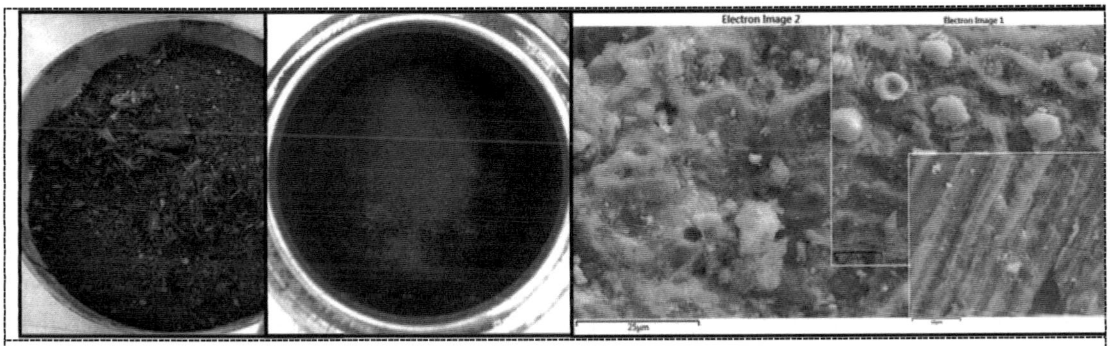

Figure 1. Palm oil boiler ash.

Table 1. Elemental composition of POBA.

Elements	Weight (%)
Carbon (C)	76 – 84
Oxygen (O)	11 – 16
Silicone (Si)	0.3 – 3.0
Potassium (K)	0.8 – 2.1

Calcium (Ca)	0.8 – 2.4
Magnesium (Mg)	0.4 – 1.1
Phosphorus (P0	0.2 – 0.5
Iron (Fe)	0.2 – 0.7
Chlorine (Cl)	0.2
Sulfur (S)	0.1 – 0.3
Aluminum (Al)	0.1 – 0.3

Table 2. Physical properties of Rediset.

Additive Properties	Index
Appearance at 25°C	Liquid
Pour point (°C)	< -20
Colour	Dark brown
Odour	Slight
Solubility in water	Partly soluble
Boiling point (°C)	215
Flash point (°C)	230
Viscosity at 40°C (mPas)	135
Density at 40°C (g/cc)	0.962

2.2. Sample preparation
The optimum dosage of Rediset recommended by the literature was 2% of base binder weight for conventional asphalt binder [24-26]. Rediset used in this study was incorporated into asphalt 60/70 as control sample namely 0%. For preparation of the modifier for this project, POBA was heated for 20 hours in the oven at 70°C according to ASTM C70-20 and sieved through a 0.15 mm (#100) sieve size. Based on the specification of the product, the blending would be applied beforehand, by adding constant percentage weight of Rediset into asphalt binder at 130°C and mixed at 5000 rpm for 2 hours. The blending of asphalt binder, with POBA and Rediset was prepared by using the shear mixed and stirrer. The blending was performed by applying a constant shear rate of 5000 rpm for two hours to reach the homogeneity of the binder mixture. The blending procedure started with a small tin can have preheated asphalt 60/70 weight of 500 grams was heated at a constant heat at 155°C before being added with 10grams of Rediset and variation weight of POBA. The mixing temperature according to the literature mixing temperature of warm mix additive specification that is 16 to 33°C less than hot mix asphalt temperature.

As for modified binder, the mixing temperature can be higher than 140°C in order to maintain fluidity of asphalt at lower temperature as well as to improve rutting resistance [27-29]. The temperature is fixed at 170±5°C during the mixing stage then POBA and Rediset were gradually added to the asphalt 60/70. The percentage of POBA in each binder investigated in this study is 0%, 3%, 5%, 7% and 9%. During the preparations of the blending, asphalt 60/70 was transferred into 500ml tin can then heated 155°C for about 10 minutes to avoid hardening during blending procedure and testing. The three material combinations were then stirred at 5000 rpm with on constant heat of 170°C. After the mixture was thoroughly blended, the can was removed from the heater and allowed to settle down at room temperature. The final asphalt binder blends were kept at room temperature, until further testing was complete, covered with a can lid.

2.3. Storage stability
The modified asphalt binder's stability when stored at high temperatures is determined using a storage stability test. The heated modified asphalt binder was put into a 25 mm diameter aluminum tube with a

IOP Conf. Series: Materials Science and Engineering 1229 (2022) 012009 doi:10.1088/1757-899X/1229/1/012009

length of 140 mm, and then heated for 48 hours at 163°C. After cooling, the aluminum tube is sliced into three pieces. Softening point test is used to evaluate the asphalt binder portions from the tube's top and bottom. Based on ASTM D36, the difference between softening points must be less than or equal to 2.5°C to pass the test.

2.4. Physical test

The consistency test is carried out to determine physical parameters whether POBA can be used to improve or maintain the characteristics of the asphalt binder. Physical tests such as penetration, softening point, ductility and viscosity were conducted according to the procedure as reported in standards ASTM D5-13, ASTM D36-14, ASTM D113-07 and ASTM D4402-15 respectively. These tests were conducted to determine the change in Asphalt 60/70 binder properties after the addition of POBA and Rediset.

2.4.1. Penetration. The penetration test determines the hardness or consistency of an asphalt binder by measuring the distance in tenths of a millimeter that a standard needle vertically penetrates a sample under known loading, temperature, and time conditions [30]. The asphalt binder sample is placed into a test container and set in a water bath for minimum two hours at a test temperature of 25°C. Thus, the container is transferred to the test equipment under a needle with a 100 gram of weight. The needle is enabled to penetrate the sample for 5 seconds and the reading is provided as a penetration value in units of 0.1 mm (dmm). The average penetration value from five reading was recorded.

2.4.2. Softening point. Asphalt binder is viscoelastic substance without sharply defining the melting points; as the temperature rises, it becomes softer and less viscous. Therefore, softening points must be determined for its consistency. The consistency of the asphalt binder starts flowing where the temperature is in between solid and liquid behaviour. The test procedure is as follows, when two plates (horizontal) of asphalt binder, are set into a shouldered brass rings which are each supported by a steel ball and heated at a controlled rate in a liquid bath. The softening points is reported as the average temperatures at which the two plates are soft enough to allow each ball, covered in asphalt binder, to fall a distance of 25 mm [31].

2.4.3. Ductility. Ductility test is used to identify the tensile properties of asphalt binder and is used to meet standard and specification requirements. When two ends of a briquette specimen are sampled, the elongation distance before breaking is used to determine the ductility of the asphalt binder. Samples was prepared in the "bone shaped" briquette mold and left to cool down to room temperature for 35±5 minutes. It was then stored it in the water bath for 35±5 min at test temperature of 25°C. Then, the sample is taken out and dried with paper towel and the excessive binder was trimmed to make the molds nicely level full. It was then arranged accordingly in a water bath at 25°C for 90±5 minutes before testing in the ductility machine. To initiate the testing, the end of binder samples is fixed and they are pulled horizontally at a uniform speed of 5cm/min until the sample breaks. Ductility values in centimeter is recorded as soon as modified asphalt binder samples start to break apart [32].

The penetration index (PI) is used to identify temperature susceptibility of asphalt binders. PI gives a possibility to predict its behavior in an application determined by the relationship shown in the following equation:

$$PI = \frac{20\,(1-25A)}{1+50A} \tag{1}$$

Where;

$$A = \frac{\log \text{Pen at } T_2 - \log \text{Pen at } T_1}{T_2 - T_1} \tag{2}$$

In equation (2), T_1 and T_2 are two different temperatures at which penetration is measured. The value *A* can also be determined from softening point temperature with assumption that all asphalt binder has a penetration of 800 at their softening point. Thus, the relationship as per following equation to determine *A* is also valid to determine PI as shown in equation (3).

$$A = \frac{\log(800) - \log(\text{Pen})}{T_{R\&B} - T_{pen}} \quad (3)$$

2.4.4. Viscosity. Viscosity is a physical characteristic of a fluid, which indicates how resistant it is to flow. The ratio between the applied shear stress and the shear strain rate is known as viscosity. To assess the flow behavior under varied circumstances, many types of viscometers are employed. The capillary viscometer, cup viscometer, and the rotational viscometer are all commonly used viscometers for asphalt binder properties [33]. The viscosity of the asphalt binders was investigated using a rotational viscometer. The characteristics of the asphalt binder blending at high temperatures were evaluated using rotational viscosity test. The viscosity is obtained by determining the torque necessary to keep a spindle immersed in a stainless-steel chamber filled with an asphalt binder sample rotating at a constant speed. The test can determine viscosity at any temperature allowing for the creation of a temperature and viscosity graph that might be utilized to calculate blending and compaction temperatures for asphalt mixture design.

In this study, temperature ranging start from 100°C to 200°C were used to determine the value of viscosity. A total of three readings was obtained at intervals of one minute with every 10°C rise. Heat was applied to the sample before it was poured into the cylinder. For each round of the test, around 10±1 gram of binder is required. A sample chamber is filled with hot asphalt binder, which is then kept in thermos container at the correct test temperature (38°C to 260°C). Thus, the spindle was lowered into the binder, and the entire assembly was left to stand until it reached the stable temperature required for testing. Once all is ready, the test is performed by spinning the spindle at 20 rpm and recording the viscosity values in mPas at expected temperature. When the desired temperature has been reached, the result was recorded.

Penetration-Viscosity Number (PVN) may also be used to measure the temperature susceptibility of asphalt binders. PVN is utilized in this study to further confirming the temperature susceptibility of the asphalt binder. For this study, the number is based on penetration at 25°C and viscosity at 135°C & 165°C. PVN is estimated by the following equation (4):

$$\text{PVN} = -1.5 \, \frac{\text{L} - \log_{10} \eta}{\text{L} - \text{M}} \quad (4)$$

Where;
L = 4.258 – 0.79674 $\log_{10}$ (PEN),
M = 3.46289 – 0.61094 $\log_{10}$ (PEN),
η = viscosity at 135°C (cSt),
PEN = Penetration at 25°C (dmm).

The susceptibility criteria considered in this study are as follows: when PVN >-0.6 binders have low temperature susceptibility and value from -1.0 to -0.6 meaning the binder have moderate temperature susceptibility. While PVN value less than <-1.0 have a high temperature susceptibility. Normally asphalt binders have a PVN value of +0.5 to -2.0, with the relationship of lower PVN indicates higher temperature susceptibility.

3. Result and Discussion

3.1. Storage stability
The result of storage stability test of modified warm mix asphalt binder in Table 3 described that the separation ratios were in the range 0.7 to 2.3 which demonstrating the heat stability of the modified

binders by POBA and Rediset. The modified asphalt binders passed the softening point criteria (the temperature differential between the bottom and top sections of test tube should be less than 2.5°C).

Table 3. Storage stability results test of all binders.

Sample	Top	Bottom	The Different
0%	48.7	51.1	2.4
3%	47.1	46.3	0.8
5%	46.7	48.7	2.0
7%	47.3	46.6	0.7
9%	48.4	46.1	2.3

3.2. Physical test

Table 4 describes the penetration value increase when Asphalt 60/70 blended with 2% weight of liquid Rediset (without POBA added), which means the hardness of the binder is increasing when Rediset is added. The penetration value is decrease when blended with 3% weight of POBA, but increased when 5% of POBA added. While the decrement of penetration value of 7% is not very much compared to 3%, the other blending with 9% showed significant difference with 72.48 dmm. This result means that by adding 5% to 7% of POBA to the asphalt binder could maintain the penetration value of the base asphalt specification.

The softening point value of the asphalt binder decreased when Asphalt 60/70 blended with 2% weight of Rediset (without POBA added), the effect of Rediset on the softening point results in a decrease of 48,40°C to 44,25°C. However, the softening point value increases when blended with 3% weight of POBA, and showed a consistency decrease when more weight of POBA (5%, 7% and 9%) added. Nonetheless, the softening points are found to be decreasing than the base binders, although the decrement of softening point value of 3% and 5% are not very much compared to 7% and 9%, this result proves that the incorporation of POBA and Rediset to the Asphalt 60/70 would reduce the softening temperature values. The sensitivity of the asphalt binders to temperature can be identified through the penetration index (PI) value. The PI value varies from around -3 for a very temperature susceptible binder to +7 for a strongly blown low temperature susceptibility. The temperature susceptibility of the asphalt binder decreases if PI value increases while the normal PI value for asphalt binder is ranging between -2 to +2. Asphalt binder that has PI value more than +2 has low susceptibility to temperature while excessively susceptible to high temperature if the PI value is less than -2. Binders containing asphalt mixtures with higher PI values (meaning binders with lower temperature susceptibility) are preferred as these asphalt mixtures have a good resistance to cracking and rutting at low and high temperatures.

Based on the results of softening point test and penetration test, we can evaluate the penetration index (PI) and temperature susceptibility of all modified asphalt binder. Table 4 shows that the penetration index of all binders is at high temperature susceptibility with ranges PI value of -1 to -3 relatively. In this research, the incorporation of Rediset to asphalt binder changed PI values to more high temperature susceptibility, but decreasing it when adding the 3% and 5% of POBA. But furthermore, adding more weight (7% and 9%) of POBA has turn the PI value increased significantly. At low temperatures, the cohesive strength of asphalt binder is characterized by force ductility.

The ductility value of the binder samples was not constant. However, generally, the ductility decreases with the addition of POBA with the 9% is the smallest value. The control asphalt binder (0%) sample have the same ductility value with base asphalt binder of 150 cm, and sample for 5% and 7% have the value above the 100 cm. Meanwhile sample for 3% and 9% having a ductility value below the 100 cm respectively, results of these ductility tests exceed the minimum length specifications required for elasticity behavior in binder properties.

In Table 5, the standard deviation value indicates the variability, and coefficient of variance shows the extent of variability. While the standard deviation values were in the allowable average limit and

IOP Conf. Series: Materials Science and Engineering 1229 (2022) 012009 doi:10.1088/1757-899X/1229/1/012009

reflect the repeatability of the results of the asphalt binders for each test, the coefficient of variation confirms the limitation of the errors.

Table 4. Summary of all physical tests on all binders.

Experimental Test	Specification	0%	3%	5%	7%	9%	Remarks
Penetration (ASTM D5)	60-70	75.28	53.44	68.16	62.94	72.48	in consistent
Softening Point (ASTM D36)	> 47	44.25	47.25	47.05	43.95	39.55	in consistent
Penetration Index	-3 to 7	-1.82	-1.76	-1.24	-2.33	-3.46	decrement
Ductility (ASTM D1113)	>150	>150	97.0	129.5	110.5	84.1	in consistent

Table 5. Summary of penetration, softening point, and ductility tests result.

Sample	Penetration (dmm)			Softening Point (°C)			Ductility at 25°C (cm)		
	Ave	SD	CV	Ave	SD	CV	Ave	SD	CV
0%	75.28	1.55	2.05	44.25	0.25	0.56	150.00	0.00	0.00
3%	53.44	1.30	2.43	47.25	0.25	0.53	97.00	3.00	3.09
5%	68.16	1.60	2.34	47.05	0.05	0.11	129.50	6.50	5.02
7%	62.94	0.85	1.36	43.95	0.15	0.34	110.50	8.50	7.69
9%	72.48	0.81	1.11	39.55	0.25	0.63	84.10	8.56	10.18

Ave = Average; SD = Standard Deviation; CV = Coefficient of Variation

3.3. Viscosity test

In this test we used rotational viscometer that is based on the idea that the torque required to spine a material in a fluid at a given speed is a function of the fluid's viscosity. Basically, this instrument consists of a thermostatically controlled chamber containing a hot asphalt binder. The spindle is lowered into the fastener and spin. The torque required to spin the spindle is determined, and converted into the viscosity of the asphalt binder (in Pas).

3.3.1. Mixing and compaction temperature. The temperature test started at 100°C and ends in 200°C, when the results are relatively constant respectively. The viscosity value ranges from 100°C to 130°C, and at the ends of the test temperature ranges more than 180°C are relatively constant. The rotational viscosity of asphalt binder is specified at 135°C or 165°C (Table 6).

Table 6. Rotational viscosity tests on all binders.

Experimental Test	Temp	0%	3%	5%	7%	9%
Rotational Viscosity (mPas)	135°C	752.33	770.83	734.33	718.17	711.67
	165°C	295.17	338.00	313.17	270.17	254.50

3.3.2. Penetration-viscosity number (PVN). Table 7 shows that the addition of POBA had reduced the resistance to temperature susceptibility. However, the pattern of reduction did not in any consistent form. This happened because the penetration values are not in consistent pattern as well. The same observation happened in Abdulrahman [29] study on 60/70 asphalt binder modified with cup lump rubber (CLR). PVN of modified binder increased with the increasing of CLR percentage. When Evotherm, a one of warm mix asphalt additive incorporated into the binder, PVN had become inconsistent with the amount of WMA.

Table 7. Summary of viscosity tests on all binders.

Experimental Test	0%	3%	5%	7%	9%	Analysis
Penetration-Viscosity Number	0.38	0.02	0.23	0.10	0.25	decrement
Activation Energy	40.16	40.01	40.25	41.73	44.17	increment
Viscosity-Temp. Susceptibility	1.8661	1.8562	1.8777	1.9515	2.0811	increment

4. Conclusion

From all the experimental test had been done, the study can be concluding as:

(a) Various POBA content in Asphalt 60/70 and Rediset gives a variation of physical and chemical characteristics. POBA as modifier to be found had an effect on warm mix asphalt binder but at certain composition it is still in accordance with the specification.

(b) The consistency of physical properties showed a good term with certain composition of POBA weight incorporated, especially in addition of 7% POBA weight. The penetration values and ductility characteristics are not in consistent pattern, while softening point temperatures is in consistent decrements pattern. The rotational viscosity of each combination had decreased with the increment of temperature even though the trend was not significantly constant. PI showed an increment with 3% and 5% and decrement pattern on 7% and 9% as well as PVN showing an inconsistent decrement with the addition of POBA.

5. References

[1] Tricia R 2013 https://www.on-sitemag.com [12 July 2021]
[2] Akhtar T M, Hassan K A, Rehman Z, Qamar S, & Niaz A M 2020 *J. Cons. and Build. Mat.* **242** 24
[3] Raman N A, Hainin M R, Hassan N A, & Ani F N 2015 *J Tek.* **72(5)** 105–110
[4] Foong S Z, Andiappan V, Tan R R, Foo D C Y, & Ng D K S 2019 *J. Proc.* **7(100)** 27
[5] Irvan 2018 *IOP Conf. Series: Mat. Sci. and Eng.* **309(012136)** 8
[6] Mahlia T M I, Abdulmuin M Z, Alamsyah T M I, & Mukhlishien D 2001 *J. Ener. Conv. And Man.* **42(18)** 2109–2118
[7] Rusbintardjo G, Hainin M R & Yusoff N I M 2013 *Cons. and Build. Mat.* **49** 702–711
[8] Mohammed B A Koting, S, Ramli S N H, Karim M R, & Mohammed A M B 2020 *J. Cons. and Build. Mat.* **262** 11
[9] Karim M R, Khandaker M U, Asaduzzaman K, Razak H A, & Yusoff S B 2017 *J. Ind. And Built Env.* **28(4)** 13
[10] Yahya Z, Kamarudin H, Abdullah M M A B, Ismail K N, & Razak A R 2016 *Key Eng. Mat.* **673 21-28**
[11] Manikam M K, Halim A A, Haafiah M M, & Krishnamoorthy R R 2019 *Desalination and Water Treat.* **149(2019)** 23-30
[12] Al Mulali M Z, Awang H, Abdul Khalil H P S, & Aljoumaily Z S 2015 *Cement and Concr. Comp.* **55** 129-138
[13] Ramadhani F, Aryanti E, & Saragih R 2015 *J. Agrotek.* **6(1)** 9-16
[14] Aslam M, Shafigh P, & Jumaat M Z 2016 *J. Cleaner Prod.* **126** 56-73
[15] Lau S Y, Phuan S L, Danquah M K & Acquah C 2019 *J. Cleaner Prod.* **230** 527-535
[16] Haron K, Mohamed A T, Halim R M, & Din A K 2008 *MPOB Info Series* 1-4
[17] Shafii M A, Durai P R, Rais N M, Latif A A 2018 *Inter. J. of Eng. And Tech.* **7(4)** 3245-3248
[18] Alamawi M Y, Khairuddin F H, Yusoff N I M, Badri K & Ceylan H (2019) *Cons. and Build. Mat.* **204(2019)** 122-131
[19] Aziz M, Rahman M, Hainin M R & Wan Abu Bakar W A 2015 *Cons. and Build. Mat.* **84(2015)** 315-319

[20] Leng Z, Gamez A, & Al-Qadi I L 2014 *J. of Mat. in Civil Eng.* **26** 304–311
[21] Varamini, S 2016 *Ph.D. Thesis, University of Waterloo, Canada*
[22] Vahora S I, & Mishra C B 2017 *Int. J. of Current Eng. and Tech.* **7(3)** 1011–1015
[23] Abd D M 2017 *Ph.D. Thesis, University of Liverpool*
[24] Habal A, & Singh D 2019 *Int. J. of Pavement Eng.* **22(4)** 467–479
[25] Kataware A V & Singh D 2017 *J. Cons. and Build. Mat.* **146** 436–444
[26] Mo L, Li X, Fang X, Huurman M, & Wu S 2012 *J. Cons. and Build. Mat* **37** 239–247
[27] Yoo M Y, Jeong S H, Park J Y, Kim N H & Kim K W 2011 *Trans. Res. Rec.* **(2208)** 40–47
[28] Kim H H, Mazumder M, Lee S J, & Lee M S 2018 *J. Traffic and Trans. Eng.* **5(3)** 197–206
[29] Abdulrahman S, Hainin M R, Idham M K, Hassan N A, Warid M N M, Yaacob H, Azman M 2019 *IOP Conf. Series: Mat. Sci. and Eng.* **527** 11
[30] ASTM D5-13 2013 *Annual Book of ASTM Standards*
[31] ASTM-D36-D36M-14 2014
[32] ASTM D 113-07 2007 *Annual Book of ASTM Standards*
[33] Hunter D R N, Self A, & Read P J 2015 *ICE Publishing*

Acknowledgments
The authors would like to express their gratitude to Universiti Kebangsaan Malaysia for the financial support under the university grant DIP-2020-003 for this work.

Engineering Properties and Slope Inventory of Clayey Soil from the Trusmadi Formation in Bundu Tuhan, Sabah

Nur Ayu Sufiah Khairul[1] **and Baba Musta**[1*]

[1]Faculty of Science and Natural Resources, Universiti Malaysia Sabah, Sabah 88400, Malaysia.
[1*]Natural Disaster Research Centre (NDRC), Universiti Malaysia Sabah, Sabah 88400, Malaysia.

Email: babamus@ums.edu.my

Abstract. The study area in Bundu Tuhan mainly consists of the metasedimentary rock of the Trusmadi Formation. Trusmadi Formation aged Palaeocene to Middle Eocene made up of interbedded dark grey shale and thin sandstone that shows the characteristics of deep marine sediment. Shale bed also known as phyllite unit is the major rock unit in Trusmadi Formation that is usually weathered and produced clayey soil. Clayey soil is often acknowledged as problematic soil due to its properties that tend to hold water highly compared to sandy soil, which will cause further engineering problems due to the soil creeping. The study area is prone to landslide activity, especially in the Trusmadi Formation's slopes. Hence, this paper aims to determine the engineering properties of clayey soil from the Trusmadi Formation. Two slopes were also selected for laboratory analysis and slope inventory. The results of the analysis show that both samples are poorly sorted materials which are classified as sandy and silty clay soil and silty clay soil. The moisture content is 25.98% to 37.10%, specific gravity ranged from 2.52 to 2.57, and plasticity index ranged from 21.68% to 23.91%. The soil has inactive to a normal type of clay (0.59 to 0.97) with a very high swelling capacity (9.29% to 11.43%). The maximum dry density (MDD) ranged from 1.44 to 1.53 and optimum moisture content (OMC) of 25.47% to 25.67%. Also, both samples are classified as impermeable soil with a K value ranging from 1.74×10^{-8} m/s to 2.45×10^{-8} m/s. While, based on the slope inventory inspection, both slopes were comprised of metagreywacke, where the first slope was an embankment slope, and the second slope was a cut slope. The first slope has poor vegetation cover which leads to moderate erosion and instability that further caused failure which is rotational debris slide. As for the second slope with average vegetation cover, there is only minor erosion and instability present on the slope, and this slope also shows slow soil movement that indicates soil creep.

Keywords: Trusmadi Formation, clayey soil, slope inventory.

1. Introduction

The study area in Bundu Tuhan is located on the West Coast of Sabah, with a total area of 100 km^2 and coordinate values of 5°57'23'' to 6°02'52'' North and 116°28'11'' to 116°33'38'' East. The study area is surrounded by a hilly topography with a maximum elevation of 432816 cm and a minimum elevation of 47244 cm.

The highland area dominates the Northeast part of the study area which covers the route to Mount Kinabalu. While the lowland area has three main rivers, namely Sungai Muau at the Southwest, Sungai Kedamaian at the North, and Sungai Kenipir at the Southeast part of the study area. Figure 1 shows the sampling station and the geological map of the study area.

The study area is underlain by two rock units which are igneous rock and metasedimentary rock. Igneous rocks that cover approximately five percent of the whole area comprised of crystalline basement aged Triassic to Cretaceous and igneous intrusion aged Late Miocene to Pliocene, while sedimentary rocks comprised of Trusmadi Formation aged Palaeocene to Middle Eocene and Crocker Formation aged Late Eocene to Early Miocene that cover approximately 40% and 55% each of the whole area [7, 12, 17].

Trusmadi Formation has undergone a long history of tectonic activity which caused the formation of metamorphic rock such as metagreywacke, slate, and phyllite. Besides, the tectonic activity also impacts softer rocks into shearing which further forms the shear zone [11]. Based on fieldwork observation of Trusmadi Formation slopes that consist of a thick layer of saprolite which is a mixture of sand, silt, and clay materials, the soil slopes have weathered into grade V1 which rocks are destroyed into residual soils.

Besides, there is also an exposure of rock slopes that consist of mudstone, metamudstone, and metasandstone that associates with quartz veins, which most of the slopes have weathered into grade IV (highly weathered rock) and grade V (completely weathered rock). Most of the slopes experienced high-grade tectonism causing extensive structures on the rock surface (Figure 2).

The landslide activities take place repeatedly in the study area, mostly at the steep slope of the Trusmadi Formation. General geological input of the Trusmadi Formation has been recorded by the previous researcher [9, 14, 16]. In consequence, the main objective of this paper is to ascertain the properties of clayey soil in accordance with the current situation by engineering properties and correlated with slope inventory data from Trusmadi Formation's slopes.

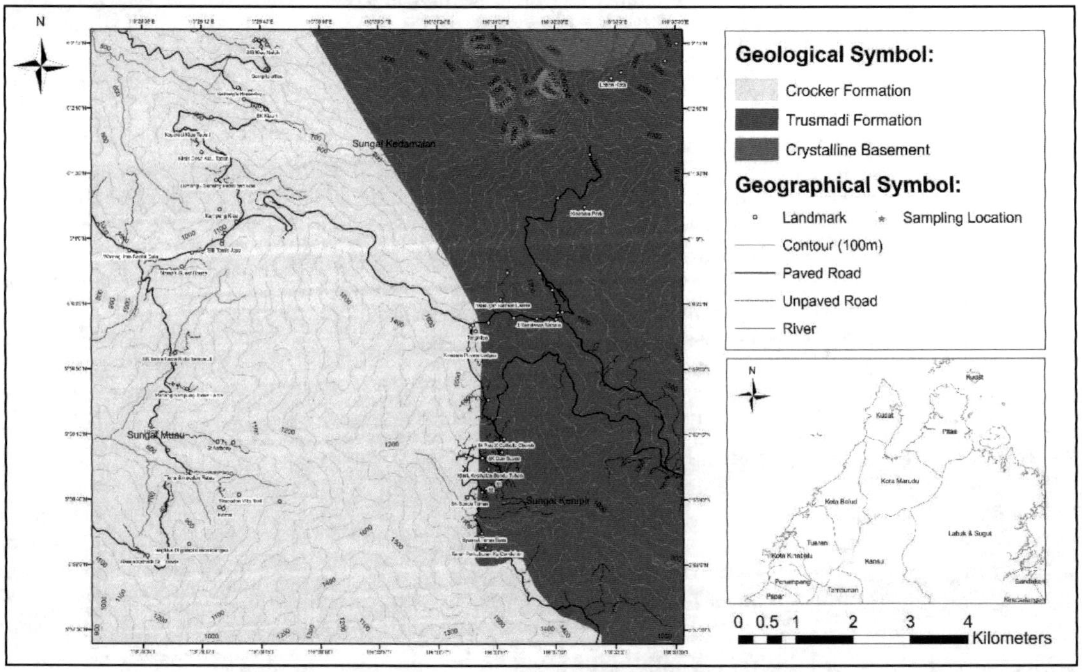

Figure 1. Study area map in Kota Belud-Kundasang, Sabah.

Figure 2. Extensive structures presence on the rock slope of Trusmadi Formation.

2. Methodology
The methodology consists of fieldwork observation, sample collection, and laboratory work. The soil samples were collected from two Trusmadi Formation slopes, namely Slope-1 (S1) and Slope-2 (S2) (Figure 1). About 30 kg of soil samples were packed in airtight plastic to preserve the natural state of moisture content before being brought to the laboratory for the Physico-chemical analysis and soil engineering analysis.

2.1. Physico-chemical analysis
Physico-chemical analysis was used to identify the basic properties of clayey soil from two Trusmadi Formation slopes. The tests included in this analysis are soil moisture content, soil organic matter content, particle size distribution, soil specific gravity, and Atterberg limits. The analysis was carried out in compliance and accordance with the methods of testing for soils for civil engineering (BS 1377-1990).

Soil moisture content was measured using the oven-dry method at 105 °C and further dried inside the furnace at 400 °C to measure the soil organic matter content. For the particle size distribution analysis, the pipette method was used to determine the percentage of silt and clay, and a dry sieve was used to measure the percentage of sand particles. The soil specific gravity is determined using the pycnometer method. The Atterberg limits, which consist of plastic limit, liquid limit, linear shrinkage, plasticity index, and clay activity, are used to determine the plasticity characteristics of the clayey soil [19].

2.2. Engineering analysis
The engineering properties consists of compaction and permeability test. Tests used are Proctor compaction test and permeability test (falling head method), which were carried out in compliance and accordance with the British Standard Code of Practise BS 1377-1990 (Methods of Test for Soils for Civil Engineering Purpose) and the American Society for Testing and Materials Code of Practise ASTM D698-2012.

Proctor compaction test used dynamic compaction method to determine MDD and OMC of a soil sample hit by 2.5 kg rammer for 27 times. Besides, the permeability test used the falling head method for fine grains sample to acquire the degree of permeability by using classification from [20], which ran for six hours in total. The results from the Proctor compaction test are required to determine the exact amount of soil sample and water needed for preparation.

2.3. Slope inventory
Slope inventory was performed to assess the physical condition of a slope either it is stable or unstable.

This analysis emphasized three main aspects such as a) slope with evidence of landslide or potentially for landslide, b) slope with a minimum height of five meters and runout distance of 10 m, and c) critical slope that affects main road and facilities including school area and village.

This analysis used the Disaster and Slope Risk Assessment (PBRC) method which has been introduced by the Department of Mineral and Geoscience (JMG) in 2015. Before conducting fieldwork observation, landslide occurrences in an area are observed by using updated orthophoto and hillshade maps from 2021 acquired from SASPlanet software. During fieldwork observation, materials needed are proforma form, study area base map, compass, and camera, where coordinate, village name, road name, and pictures showing the landslide and its surrounding are required. The proforma form is the most important thing to be completed to achieve the objective of this analysis, which consists of four categories which are a) proforma slope inventory that will be used when the critical slope is present to describe the physical condition of the slope, b) incident proforma that will be used when failures occur to describe detailed information of the landslide, c) structural proforma that will be used to describe the current condition of the structures present inside the buffer zone of the critical slope, and d) geology proforma that will be used to record rock slope data to further determine the mode of failure (Table 1).

Table 1. Characteristic and usage of each proforma form.

Proforma	Required data	Usage
Primary Slope Inventory	Lithology, feature type, slope dimensions (length, height, width, and angle), slope cover, and drainage	Used when critical slope is found (with or without landslide) to record slope data in the study area
Incident	Erosion and instability, and failure type	Used when landslide (active or remnant) is found on the critical slope to record its physical properties
Structural	Existence and condition of facility	Used when facilities (road or/and building) are found near the critical slope to check whether it was affected or not
Geology	Slope geometry, geotechnical unit and thickness, discontinuities, and mode of failures	Used when rock slope with 30% exposures is found (with or without failure) to record structural geology data

3. Results and Discussion

3.1. Engineering properties

Table 2 presents detailed results from the Physico-chemical and engineering analysis of clayey soil from Trusmadi Formation slopes. Moisture content is a ratio between the weight of water in the sample and the weight of solid, which is often affected by organic matter content, soil texture and structure, and bulk density [2]. The moisture content values at 25.98% to 37.10% are classified as wet to very wet soil [15]. While the organic matter content values at 1.94 to 2.01 classified the soil as mineral soils that indicate low organic matter content [10].

Based on the percentage of sand, silt, and clay, soil samples from Slope-1 (S1) and Slope-2 (S2) are classified as well-graded and poorly sorted materials of sandy and silty clay soil and silty clay soil each [8]. Specific gravity is directly proportional to the density of soil, with values of 2.52 to 2.57. These values can be affected by the presence of plant roots and porous particles which will decrease the density of the soil [13].

Atterberg limits of cohesive soil can exist in any four states such as solid, semi-solid, plastic, and liquid depending on the moisture content and type and amount of clay mineral content. The boundary between each state is known as shrinkage limit, plastic limit, and liquid limit [3]. Soil samples from Slope-1 (S1) and Slope-2 (S2) are classified into high plasticity material with plasticity index values ranging from 21.68% to 23.91% (Figure 3) [19].

Clayey soil is also referred as a material with slow swelling and shrinkage behaviours, which are affected by differences in the period of precipitation and evapotranspiration, poor surface drainage, moisture content, and clay mineral content [3]. The linear shrinkage values at 9.29% to 11.43% are classified as poor-quality soil in soil mechanics [6]. The linear shrinkage values are controlled by the clay activity, in which values ranging from 0.59 to 0.97 are classified as inactive to a normal type of clay [18].

Proctor compaction test applied mechanical energy to a soil to rearrange its particles and reduce the void ratio, which can be useful to increase soil shear strength and bearing capacity, decrease subsequent settlement under working loads, and decrease soil permeability to block water flow in soil mechanics. This test provides two significant data of OMC values at 25.47% to 25.67% and MDD values at 1.44 to 1.53 which means the soil in the area should be compacted at calculated OMC and MDD for maximum strength in any engineering work [2].

Besides, permeability refers to the rate at which water flows through a material under the influence of a hydraulic gradient, which is also affected linearly by porosity. The K value from the permeability test ranged from 1.74×10^{-8} m/s to 2.45×10^{-8} m/s is classified as impermeable [20], which will further cause engineering problems such as waterlogging.

Table 2. Physico-chemical and engineering properties of a clayey soil from Trusmadi Formation.

Slope no.	**Slope-1 (S1)**	**Slope-2 (S2)**
Weathering grade	VI	VI
Moisture content (%)	25.98	37.10
Organic matter content (%)	1.94	2.01
Sand (%)	11.20	11.20
Silt (%)	64.22	51.89
Clay (%)	24.57	36.91
Specific gravity	2.57	2.52
Plastic limit (%)	31.72	31.47
Liquid limit (%)	55.64	53.15
Shrinkage limit (%)	9.29	11.43
Plasticity index (%)	23.91	21.68
Clay activity	0.97	0.59
Optimum moisture content, OMC (%)	25.67	25.47
Maximum dry density, MDD (Mg/m^3)	1.53	1.44
Permeability (m/s)	2.45×10^{-8}	1.74×10^{-8}

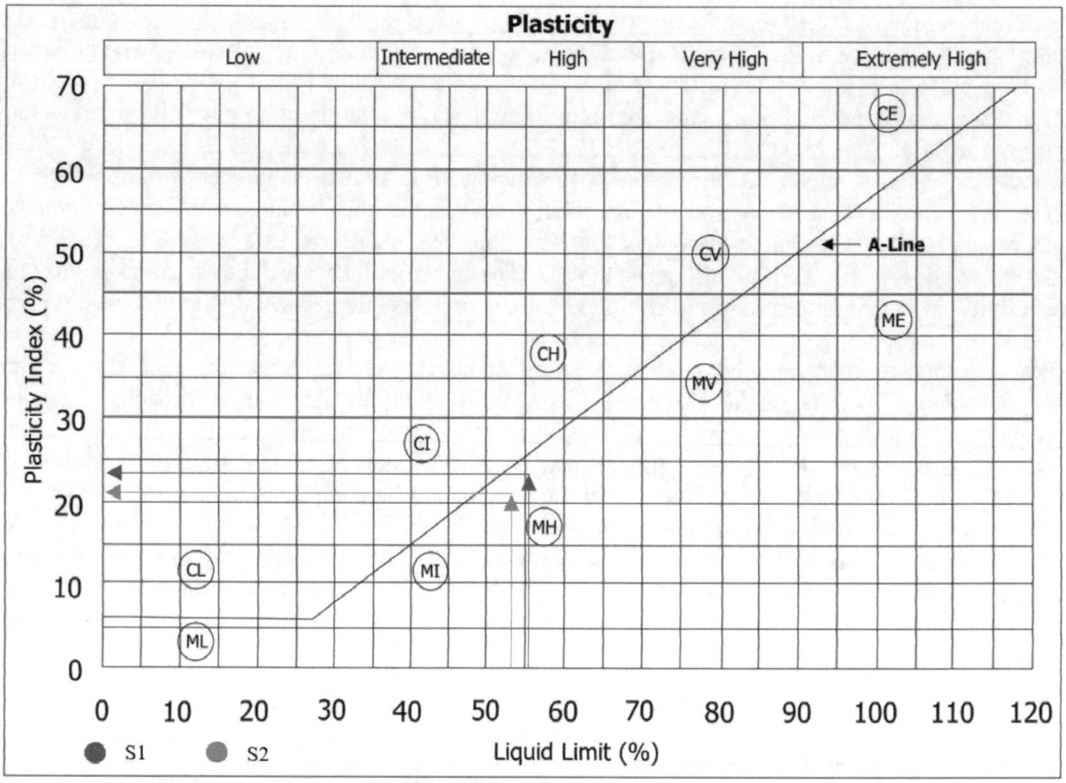

Figure 3. Plasticity chart of clayey soils from Trusmadi Formation slopes.

3.2. Slope inventory

Table 3 shows the slope inventory data acquired from fieldwork observation. It should be mentioned that the proforma forms are filled out according to the slope physical condition, which means both are soil slopes. Hence, the geology proforma is not used in this study.

The Slope-1 (S1) has a feature type of embankment slope that was further cut to a single terrace recently. This slope has dimensions of 33 m length, seven (7) meters height, 51 m width, and 35° gradient angle, which picture a slope with a high possibility of landslide occurrence due to the height that more than five (5) meters and very steep gradient angle [5]. The slope has a poor vegetation cover of around 35%, which will cause more exposure to weathering and erosion. The slope has no drainage at the bottom and top, but it has some seepage of water that flows out of the slope due to the small pipe that exists at the top. A lack of drainage systems will cause the slope to get more exposure from upstream runoff, which will lead to water irrigation problems. Based on the data acquired from the incident proforma, the slope has moderate erosion and instability, such as the presence of sheet, rill, mass movement, and fretting at some parts of the slope. Erosion can worsen on the slope without any covers, which further leads to failure, such as the rotational debris slide on the Slope-1 (S1) (Figure 4a). Sustainable human safety needs to be considered for facilities such as roads and buildings. Paved road networks are found at both the top and bottom of the slope known as Jalan Tenompok, which was the main road for the villagers in that area. Besides, there is a house at the bottom with a 10 m distance from the foot slope. The road and the house are still in a good condition without any impact from the failure that happened on the slope since the landslide is considered small and localized. Based on the eyewitness, the landslide was the most recent one that failed a week after heavy rainfall. The terrace was made to stabilize the slope and reduce any possible risk regarding landslide activity.

The Slope-2 (S2) has a feature type of cut slope with dimensions of 76 m length, 16 m height, 122 m width, and 11° gradient angle, which shows the slow movement of landslide activity. The slope was covered with grass, shrubs, and trees for around 80%, which is average. Roadside drain is found at the bottom of the slope where the condition is fair with no blockage. Since the slope has been covered more than half, the presence of erosion and instability such as sheet, fretting, and mass movement is minor and exposed only at the bottom part of the slope. Failure type for this slope was classified as soil creep, which can be recognized by uneven surface, tilted tree, and crack on the road (Figure 4b and Figure 5a). This slope has been cut due to the road construction to facilitate networks among the villagers. An average condition of paved road is found at the bottom of the slope with the presence of longitudinal crack extending to 0.99 m (Figure 5a). There are also houses present at the top and bottom of the slope with a distance range from zero (0) meter to 10 m. Three houses located at the top and one house located at the bottom of the slope are still in a good condition, while another two houses located at the bottom of the slope are slightly damaged which may be triggered by the soil movement from the upper slope (Figure 5b).

Table 3. Slope inventory data of a clayey soil from Trusmadi Formation.

Proforma	Data	Slope-1 (S1)	Slope-2 (S2)
Primary Slope Inventory	Lithology	Meta-greywacke	
	Feature type	Embankment	Cut
	Length (m)	33	76
	Height (m)	7	16
	Width (m)	51	122
	Angle (°)	35	11
	Slope cover	Poor	Average
	Drainage	No drainage	Roadside drain
Incident	Erosion and instability	Moderate	Minor
	Failure type	Rotational debris slide	Soil creep
Structural (Facilities)	Existence	Top and bottom	Top and bottom
	Condition	Good	Average

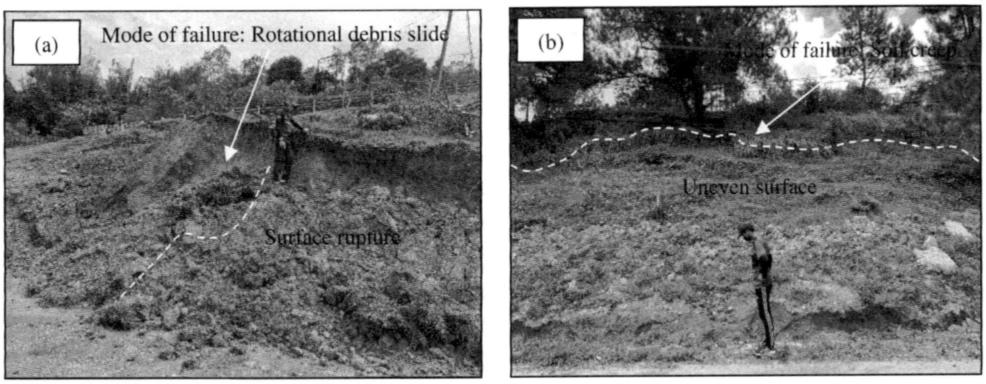

Figure 4. (a) Rotational debris slide on Slope-1 (S1) and (b) soil creep on Slope-2 (S2).

Figure 5. (a) Longitudinal crack on the main road near Slope-2 (S2) (b) Slightly damaged houses.

4. Conclusion

The engineering properties and slope inventory enabled us to acquire the properties of clayey soil from the Trusmadi Formation slopes. The results of the analysis show that the study area is dominated by clayey soil from Trusmadi Formation. The soil has a high percentage of water content, low specific gravity which also indicates less dense soil, high plasticity index with high shrinkage capacity, and is categorized as impermeable soil which will lead to further engineering problems such as waterlogging. Subsequently, this problem will cause the slope to become unstable which contributed to the occurrence of landslides such as rotational debris slide and soil creep. This indicates that the soil is not suitable for heavy engineering construction. But any construction work to be carried out is still considerable since Trusmadi Formation is the major rock unit in the study area, on condition appropriate mitigation measures are practiced ensuring the stability of the soil slope. It is proposed that the soil needs to be compacted to OMC values at 25.47% to 25.67% and MDD values at 1.44 to 1.53 as these is the range within which the soil has maximum strength.

References

[1] ASTM D698 2012 *Standard Test Methods for Laboratory Compaction Characteristics of Soil Using Standard Effort (12 400 ft-lbf/ft^3 (600 kN-m/m^3))* (West Conshohocken, Pennsylvania: ASTM International)

[2] Balasubramanian A 2017 *Characteristics of Soil Profile* (University of Mysore, Mysore: Centre for Advanced Studies in Earth Science)

[3] Bell F G, Cripps J C and Culshaw M G 1986 A review of the engineering behaviour of soils and rocks with respect to groundwater *Geological Society Engineering Geology Special Publication* **3**(3) pp 1–23

[4] BS1377 1990 *British Standards (Soils for Civil Engineering Purposes)*

[5] Ercanoglu M, Gokceolgu C and Van Asch ThWJ 2004 Landslide susceptibility zoning Yenice (NW Turkey) by multivariate statistical techniques *Nat Hazards* **32** pp 1-23

[6] Falowo O O, Olabisi W K and Olaifa G E 2017 Geochemical and Geotechnical Appraisal of the Regolith along Ado-Ilawe Road, Ekiti State, Nigeria for Engineering Construction *International Journal of Advance Scientific Research and Engineering Trends* **2**(11) pp 277-284

[7] Hall R, Cottam M, Suggate S, Tongkul F, Sperber C and Batt G 2008 *THE GEOLOGY OF MOUNT KINABALU* Publication No. 13 (Kota Kinabalu, Sabah: Sabah Parks)

[8] Head K H 1982 *Manual of Soil Laboratory Testing* 2nd Ed (London: Pentec Press)

[9] Hennie Fitria Wulandary Soehady Erfen, Asvirja Grace Gansau and William Agan Henry 2016 Engineering Properties of Soil from Unstable Slopes in Ranau-Kundasang, Sabah, Malaysia *Transactions on Science and Technology* **3**(3) pp 495-500

[10] Huang P T, Patel M and Bobet A 2008 *FHWA/IN/JTRP-2008/2 Classification of Organic Soils* (Purdue University, West Lafayette, Indiana: School of Civil Engineering)
[11] Ismail Abd. Rahim and Baba Musta 2018 The Stability of Metasedimentary Rock in Ranau, Sabah, Malaysia *Indonesian Journal on Geoscience* **5**(1) pp 23–31
[12] Jacobson G 1970 Gunung Kinabalu area, Sabah, Malaysia *Geological Survey of Malaysia, Report 8*
[13] Bowles J E 2012 *Engineering Properties of Soils and their Measurements* 4th Ed (New Delhi: McGraw Hill Education (India) Private Limited)
[14] Noran Nabila Nor Azlan, Norbert Simon, Azimah Hussin, Rodeano Roslee and Lee Khai Ern 2017 Chemical Properties Characterization of Failed Slopes Along the Ranau-Tambunan Road, Sabah, Malaysia *Sains Malaysiana* **46**(6) pp 867-877
[15] Omar H 2010 *Slope Stability Using Remote Sensing and Geographic Information System Along Karak Highway, Malaysia* (Universiti Teknologi Malaysia, Johor: Master of Science Dissertation)
[16] Rodeano Roslee, Sanudin Hj. Tahir, Baba Musta and Shariff AK Omang 2010 Geological Inputs for Landslide Hazard Identification (LHI) in the Trusmadi Formation Slopes, Sabah, Malaysia *Borneo science* **26** pp 37-51
[17] Sanudin Hj. Tahir and Baba Musta 2007 *Pengenalan Kepada Stratigrafi* (Kota Kinabalu, Sabah: Penerbit Universiti Malaysia Sabah)
[18] Skempton A W 1953 The Colloidal Activity of Clays *Proceedings of the 3rd International Conference of Soil Mechanics and Foundation Engineering, Switzerland* **1** pp 47-61
[19] Sowers G F 1979 *Introductory Soil Mechanics and Foundation* 4th Ed (New York: Macmillan)
[20] Terzaghi K and Ralph B P 1948 *Soil Mechanics in Engineering Practice* (New York: John Wiley and Son, Inc.)

Acknowledgments
A word of appreciation to Universiti Malaysia Sabah for the easy access to laboratories and research equipment, as well financed all the cost of this research under SDN scheme (SDN0037-2019).

A Review of Eggshell Powder and Fly Ash-Based Geopolymers Concrete

I.C. Moison[1], L. Gungat[1], H. Asrah[1], F.H. Chiew[2]

[1] Faculty of Engineering, Universiti Malaysia Sabah, Sabah 88400, Malaysia.
[2] Faculty of Civil Engineering, Universiti Teknologi MARA Sarawak, 94300 Kota Samarahan Malaysia

* Corresponding author: lillian@ums.edu.my

Abstract. This is a review of the current research on the development of geopolymer based of eggshell powder (ESP) and fly ash (FA). This paper observes the mechanical properties which are the compressive strength and tensile strength of eggshell powder (ESP) and fly ash (FA) based geopolymer that is formed by an alkali activation method. It is shown that the mechanical properties of alkaline activated geopolymer is significantly affected by its proportion of raw material in concrete; specifically, the fly ash (FA) to eggshell powder (ESP) ratio and the alkali activator ratio. Moreover, it is shown that curing temperature also may be a very influential factor towards the mechanical properties of the concrete. However, more experimental studies are required for eggshell powder (ESP) and fly ash (FA) based geopolymer as there are not many studies yet in this field have been conducted till date.

Keywords: fly ash; eggshell powder; alkali activation; geopolymer; mechanical properties

1. Introduction

Nowadays, one of the top global concerns is the environmental issues. One of the many environmental issues that are frequently mentioned is the emission of greenhouse gases to the atmosphere, particularly the carbon dioxide (CO_2). The most widely used construction material is Ordinary Portland Cement (OPC) which produces a huge quantity of carbon dioxide into the atmosphere. It is reported that, approximately 900kg of carbon dioxide were emitted for every 1000kg of cement production [1]. Besides that, Solid Waste Management in Malaysia has stated that one of the main environmental problems in Malaysia is poultry waste which can be regarded as a portion of solid waste. Based on World Data Atlas, production of eggs in 2019 for Malaysia was 848,658 tonnes. The high consumption of eggs in Malaysia generates big quantities of eggshells to the surrounding environment as domestic waste. Disposal of these eggshells is a big challenge as they will attract vermin and cause health issues if they are send to landfills [1]. Eggshell in landfills produce odours and promote microbial growth as they biodegrade [2].

Therefore, attempts have been made to encourage the utilization of waste material or by-product to reduce the impact towards the environment. Previous researchers have given the effort to create an alternate cementitious binder that is developed by using alkali activation method which is known as geopolymer. Eggshells contains the same amount of calcium oxide as cement, can be an excellent substitution of lime in cement production [3]. Eggshell can be crushed into powder state and can be used in developing geopolymer concrete. Eggshell also comprises of a chemical composition to the cement that as a fragmentary substitution of cement [4]. Fly ash is an industrial waste or known as by-product which can be obtain from coal combustion plant. Fly ash has the pozzolanic property that can be used as a replacement of cement. Moreover, essential minerals for geopolymerization process such as alumina and silica can be found in fly ash [3], [5]. The abundant sources of eggshell and fly ash in this country creates the convenience to utilise by-product materials as a substitution of cement in concrete. This review paper focuses on journal paper that have considered eggshell powder and fly ash geopolymer concrete. This paper provides review of research in the compressive strength and tensile strength of eggshell powder and fly ash-based geopolymer concrete.

2. Fly Ash

Fly ash is a material that possess no cementitious properties, but it will when mixed with high alkali solution. Dissolution of alumina and silica from its precursor material by high alkali solution and polycondensation reaction is called as geopolymer [5], [7]. Fly ash as an alumino-silicate precursor can be used as a replacement of cement.

2.1. Composition of fly ash

The behavior of the fly ash in concrete is highly impacted by its mineralogical, chemical and physical properties [6]. Fly ash is classified by its chemical contains. Table 1 is one of the standards to decide the type of fly ash based on its source of origin and compositions.

Table 1. ASTM Specification for Fly Ash

	Chemical Requirement	Description in ASTM C 618
Class F	$SiO_2 + Al_2O_3 + Fe_2O_3 \geq$ 70%	Fly ash normally produced from burning anthracite or bituminous coal that meets the applicable requirement for this class as given herein. This class of fly ash has pozzolanic properties.
Class C	$SiO_2 + Al_2O_3 + Fe_2O_3 \geq$ 50%	Fly ash normally produced from lignite or sub-bituminous coal that meets the applicable requirements for this class as given herin. This class of fly ash, in addition to having pozzolanic properties, also has some cementitious properties. Note: Some Class C fly ashes may contain lime contents higher than 10%.

Class F fly ash which is known as low calcium fly ash has been used vastly in producing geopolymer by alkali activation method [7]–[12]. Class F fly ash that contains low calcium contents but containing more of alumino-silicate materials that are more useful in geopolymerization. Water will be produced as products from the chemical reaction which will not interfere in the chemical reactions but simply provides the workability of the mixture during the handling. Chemical reaction in geopolymer is contrast to the chemical reaction of water in cement during the hydration process [13], [14].

3. Eggshell Powder

Calcium-rich ESP is used as precursor in fly ash based (FA) geopolymer is an innovative approach which has proven by many researchers that calcium content can contribute to an early compressive strength development. Previous studies have proven that the addition of calcium-rich precursor could by any chance boost the overall mechanical strength of the geopolymer [15]. This is because of calcium-rich precursor such as eggshell powder and marble powder could form calcium silicate hydrate (C-S-H), calcium-sodium aluminosilicate hydrate (C-(N)-A-S-H), calcium aluminate silicate hydrate (C-A-S-H) and sodium aluminate silicate hydrate (N-A-S-H) [3], [5], [7], [16], [17].

4. Alkali Activator

Creating geopolymer cement requires an alumina-silicate precursor, alkaline reagent and water [18]. Previous studies proves that the one of the factors that affecting the mechanical properties of geopolymer concrete is the alkali activator content in terms of type of alkali activator, molarity of sodium hydroxide and activator ratio [3], [5], [7], [8], [11], [14], [19]–[21].

4.1. Type of alkali activators

Soluble alkali metals are often Potassium or Sodium based, that are made to become alkaline liquids. Sodium hydroxide (NaOH) or potassium hydroxide (KOH) with sodium silicate (Na_2SiO_3) or potassium silicate (K_2O_3Si) are typically the alkaline liquid utilized in geopolymerization process [3], [5], [7], [8], [14], [20]. Because of the affordable price of sodium-based solutions, it is often used by consumer as compared to potassium-based solutions [14]. Previously, it has been proven that when using a single alkaline activator and at the same time using sodium hydroxide (NaOH), produce higher strength and faster activation as compared to potassium hydroxide (KOH) [11]. Another single alkali comparison between the uses of sodium silicate (Na_2SiO_3) and sodium hydroxide (NaOH), it was found that when forming geopolymer with sodium hydroxide (NaOH), the strength properties were firmer as compared to forming geopolymer with sodium silicate (Na_2SiO_3) [8]. It can be seen from Table 2 that when the alkali solution comprises soluble silicates (sodium or potassium silicate), the reactions take place at higher rates and leads to an improvement of strength [11], [19]. Mostly in recent research, sodium silicate (Na_2SiO_3) and sodium hydroxide (NaOH) were the preferred combination and was often opt for as the alkaline liquid as they are more economical yet giving the same effect as the potassium based alkaline [3], [5], [7], [8], [14], [20], [22].

Table 2. Compressive strength (MPa) of prisms made with a activator/fly ash ratio of 0.25 [11]

Activator		Curing temperature (ºC)	Time of curing		
			2 h	5 h	24 h
Solution 1	Sodium Hydroxide, (NaOH) 12M	65	0.0	0.0	21.2
		85	9.3	22.0	34.6
Solution 2	Potassium Hydroxide, (KOH) 18M	65	0.0	0.0	8.7
		85	1.4	9.4	23.3
Solution 3	NaOH (pellets) + sodium silicate	65	4.3	31.7	52.7
		85	39.8	48.2	54.5
Solution 4	KOH (pellets) + potassium silicate	65	0.0	9.5	38.7
		85	7.7	34.3	63.0

The SiO_2/Na_2O total ratio in solution 3 was 1.23. The SiO_2/K_2O total ratio in solution 4 was 0.63.

4.2. Molarity of sodium hydroxides solution

Higher mechanical strength of geopolymer concrete is obtain by using higher concentration (in molarity) of sodium hydroxide solution. Many studies have proven that increasing in sodium hydroxides concentration will increase the strength of the geopolymer concrete [5], [9], [12], [14]. This is because increasing in alkalinity of the alkali solution, more impairment of the glassy chain at the plane of the fly ash which triggered the reaction of the internal silicon (Si) and aluminium (Al) to

IOP Conf. Series: Materials Science and Engineering 1229 (2022) 012011 doi:10.1088/1757-899X/1229/1/012011

form gel phases which provide structural strength can be induced [23]. Table 3 shows that the compressive strength at 28 days of NaOH molarity 9M have been improved by more than 1.7 times compared to that at 6 M. However, the sample of 9 M and 12 M exhibits practically equivalent strengths [12]. Literatures also stated that molarity of hydroxides should be in an optimal amount. Excess in hydroxide ions hinders the polycondensation process in geopolymerization which in turn, resulting decrease in strength of geopolymer concrete [11], [17],[19],[24],[15]. Previous researchers have shown that the ideal concentration of sodium hydroxide solution is 10M and it should be limited around 12.5 molars since higher concentration will result in poor workability hence, will involve in lower strength [20][17]. Figure 1 shows the compressive strength of geopolymer concrete will reach its maximum value when the molarity of NaOH is increased up to the value of 10 and increasing the molarity of NaOH over time, will decrease the compressive strength.

Table 3. Compressive strength (MPa) with molarity of NaOH [12]

Specimen	Curing Temperature (°C)	Molarity of NaOH	Compressive strength (MPa)		
			14 days	28 days	91 days
FASH6SS50	60	6 M	18.4	25.9	28.2
FASH9SS50	60	9 M	27.8	43.1	47.3
FASH12SS50	60	12 M	42.9	44.8	47.5

FA: fly ash, SH: sodium hydroxide (NaOH), SS: sodium silicate

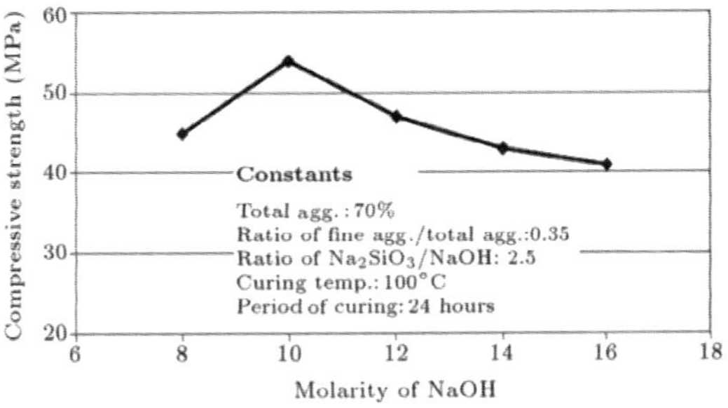

Note: Variation on individual test results from mean-minimum 0.6% and maximum 6.6%

Figure 1. Compressive strength of geopolymer concrete at 7th day with molarity of NaOH [20].

4.3. Activator Ratio (Na_2SiO_3/ NaOH)

The compressive strength of alkali-activated fly ash concrete increased with the mass ratio of sodium silicate to sodium hydroxide solution. Higher the ratio of sodium silicate to sodium hydroxide by mass, higher the compressive strength of the geopolymer concrete [3], [14], [20], [24]. However, literatures also stated that the compressive strength of geopolymer concrete increase with the activator ratio until the ration reached 2.5, afterwards it involves in decrease of strength [23][20][24]. Table 4 shows that there are increment of compressive strength until activator ratio of 2.5 and reduction in compressive strength as it reaches the ratio of 3.0. This phenomenon is because the conversion in microstructure of geopolymer, affected by the amount of sodium silicate will increase the compressive strength. While, if the amount of sodium hydroxide is insufficient to finalize the dissolution process during the formation of geopolymer and the ratios of sodium silicate to sodium hydroxide is high, this will cause the compressive strength to decrease [20], [25], [26].

Table 4. Compressive strength of sodium silicate to sodium hydroxide ratio [24].

Specimen	Solid to liquid ratio	Molarity NaOH	Activator ratio ($Na_2SiO_3/NaOH$)	Compressive Strength (MPa) At 7th days curing in room temperature
12 M 1.5 2.0	2.0	12 M	1.5	43.83
12 M 2.0 2.0	2.0	12 M	2.0	57.33
12 M 2.5 2.0	2.0	12 M	2.5	65.09
12 M 3.0 2.0	2.0	12 M	3.0	49.15

5. Curing Temperature and Curing Duration

Curing temperature is one of the notable components that affecting the mechanical strength of geopolymer concrete. Geopolymerization and a well-reacted geopolymer structure can be produce in the presence of heat as heat accelerates the chemical reaction[5][15]. From Figure 2, it shows that the compressive strength will reach its maximum value when the curing temperature reaches 100 ℃ and decreases beyond 100 ℃. The decreasing of compressive strength that surpassed 100 ℃ is because the excessive evaporation of moisture from the sample [20]. Curing at a higher temperature caused an instant loss of moisture that creating voids. Thus, the samples will fail to create a dense geopolymer because of the porous structure which will resulted in loss of strength [5][20][15]. Although, when increasing curing temperature at 75℃ to 90℃ the strength gain became less noticeable. Meanwhile, longer curing duration generated higher average of compressive strength of the geopolymer. Table 2 shows the increment of compressive strength of the geopolymer as the curing duration increases regardless of the curing temperature. Longer curing duration improved the polymerisation process [14]. However, from Figure 3, the improvement of strength became less noticeable when the curing duration surpassed 48 hours [20].

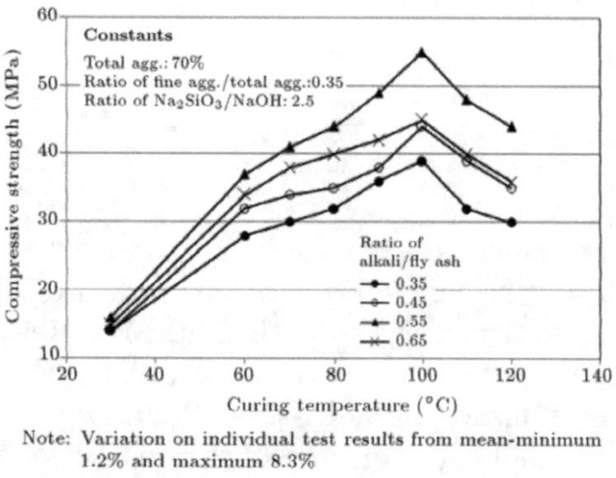

Figure 2. Compressive strength at 7th day with curing temperature [20].

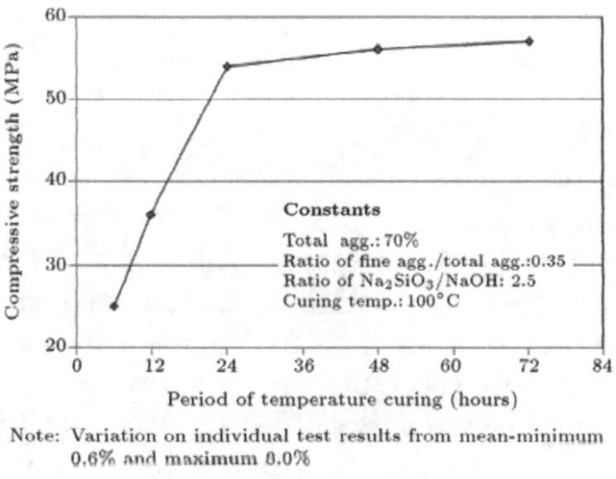

Figure 3. Compressive strength at 7th day with duration of curing [20].

6. Preparation of Geopolymer Concrete

This section will explain briefly regarding the process that has been conducted by previous researchers. Variable parameters and data obtained can vary according to their research purposes.

6.1. Materials

Dry materials comprise of fly ash, eggshell powder, and aggregates. Eggshells were collected from local vendors such as restaurants, bakeries, and poultry farms. The eggshells were rinsed and cleaned and then were dried under the sun for a day. After drying, the eggshell was ground into a powder form. Fly ash was procured from local power plant as by-product. Coarse aggregate and fine aggregate are typical crushed granite and sand. Liquid component which is alkaline liquid used in this field are typically the mixture of sodium silicate solution (Na_2SiO_3) and sodium hydroxide (NaOH). These alkaline can be obtained from local chemical suppliers. The sodium hydroxide solids were in flakes or pellets form, need to be diluted in water depending on the concentration of the solution expressed in term of molar, M.

6.2 Mixing, casting, curing, and testing

Mix sodium silicate solution and sodium hydroxide solution together at least one day prior to adding the liquid to the dry materials. Mix all dry materials in the pan mixer for about 3 minutes, then add the liquid component and continue the wet mixing for another 4 minutes. Super plasticiser is optional to improve the workability of the fresh geopolymer concrete which is usually a naphthalene sulphonate super plasticiser. After mixing, fresh fly ash and eggshell powder geopolymer concrete then compacted into mould for it to be hardened and ready for testing which are usually the compressive and tensile strength tests. The prepared samples are wrapped within the vinyl sheet to prevent moisture loss and the cured at ambient room temperature or elevated temperature using steam curing or oven dry. Most common test used to determine the mechanical properties of the geopolymer concrete is compressive strength test and tensile test.

7. Eggshell Powder and Fly Ash

7.1. Compressive strength and Tensile strength

Mechanical properties are an important indicator of the toughness of concrete to be used, either it is compatible for structural element or not. By obeying all the literatures that have been reviewed in the subtopic above, when it comes to the utilization of eggshell powder and fly ash, the proportion of these two precursors is important in determining the mechanical strength of the geopolymer. Too much of eggshell powder in a sample could lead to a porous surface and lower strength [5]. This is because of the bigger particle size of the eggshell powder compared to fly ash and the limited amount of aluminosilicate gel available for the geopolymerization process [5]. Greater moisture content is required when there is excessive amount of fly ash in a sample, this is because of the large surface area of the finer precursor particles. Other than that, balanced proportions of eggshell powder and fly ash will produce a denser structure in which a demonstration for a greater mechanical strength structure. This is because of the finer fly ash particles filled the voids produced by the larger eggshell powder particles [5].

Previous literature suggested that the optimum activator content for eggshell powder and fly ash-based geopolymer is 2.0 with the molarity of sodium hydroxide of 10 M [3], [7][20], [23]. It implies that compressive strength of eggshell powder and fly ash based geopolymer simultaneously increases with an increment in fly ash content and a decrement in eggshell powder content [3], [5], [7]. Both compressive strength of geopolymer and temperature will increase at the same time but at a certain duration. When curing at 80°C for a longer duration of 56 days, the heat produced will led to an immediate shrinkage cracks and loss of moisture which in turn resulting in lower compressive strength

[7][15]. Researches has proven that to achieve a higher compressive strength of geopolymer, it requires prolonged heat curing but for a certain range of temperatures [7][15]. Similar to the compressive strength, tensile strength geopolymer increases with the increase in fly ash content due to the higher SiO_2 content in fly ash that produces a higher of geopolymerization reaction [3], [7]. Tensile strength increases with the increment of activator ratio (Na_2SiO_3/NaOH) but decreases in higher temperature for a longer duration. Activator content of 2.0 led to a higher tensile strength. Similarly, with the compressive strength at higher temperature, heat curing for longer duration resulted to the loss of tensile strength [3], [7].

7.2. Durability

To have a sustainable concrete structure, durability is key. The cyclic wet-dry test is one of the methods used to measure the durability of concrete in a non-frost area. Previous researchers have measured the durability of eggshell powder and fly ash based geopolymer by conducting wetting-drying cycles test [15]. From the research, activator ratio of 2.0 and heat curing temperature of 50ºC can be regarded as optimum as the samples in this particularly activator ratio and curing temperature have provided the lowest strength loss among other samples with different activator ratio and temperature [15]. Similarly, in previous literatures, research on durability of high calcium clay with fly ash geopolymer versus sulfate attack was carried out and it was discovered that clay-fly ash geopolymer is a lot durable as compared to the composite of clay-cement [27]. There is no noticeable reaction in the microstructure of clay-FA geopolymer after the exposure of the sulfate solution.

7.3. Application

Fly ash and eggshell powder geopolymer can be apply in the pavement construction. Previous literatures has shown that the geopolymer fulfill the UCS requirements of MORD 2014 for rural road construction as base, subbase or subgrade material [3], [7]. Moreover, previous literatures have proven that eggshell powder and fly ash solely have lesser specific gravities, thus both materials can be utilized as lightweight fill material or backfill material for retaining wall [3], [7], [15]. For field application, dried precursor materials to be mixed on site with the optimum activator content and then spread and rolled over for pavement base or subgrade construction. Previous researchers has shown that the eggshell powder and fly ash based geopolymer can attain a satisfactory compressive strength and tensile strength at room temperature [3]. Therefore, in a tropical country such as Malaysia which the average temperature of 30ºC, cementation regularly can occur in the field and the geopolymer can be performed successfully.

Table 5. Summary of Eggshell powder and Fly ash based Geopolymer application.

Description	Application	Reference
Satisfy the weight loss and strength loss criteria specified by IRC (Indian Road Congress, 2018) after durability test.	Landfill for pavement construction material.	Durability analysis of eggshell powder-flyash geopolymer composite subjected to wetting-drying cycles. [15]
Meets the unconfined compressive strength requirements of MORD 2014 (Ministry of Rural Development)	Base, Sub-base, Subgrade	Potential Application of Heat Cured Eggshell Powder and Flyash-Based Geopolymer in Pavement Construction. [7]
Meets the unconfined compressive strength requirements of MORD 2014 (Ministry of Rural Development)	Sub-base, Subgrade	Strength Behavior of Alkaline Activated Eggshell Powder and Flyash Geopolymer Cured at Ambient Temperature. [3]

8. Conclusion

This extensive review of research data has led to the following conclusions. Firstly, the combination of alkali solution with soluble silicates gives a better result in strength improvement because of the higher rates of reaction occur when hydroxides are used with soluble silicates as the activators. Secondly, the higher molarity of alkali solution resulted in greater mechanical strength because the higher alkalinity caused an increment breakage of glassy chain at the surface of the fly ash and induced internal reaction of silicon (Si) and aluminium (Al) to provide structural strength. However, excessive in alkali concentration will result in deterioration of geopolymer strength. Thirdly, the higher the sodium silicate (Na_2SiO_3) to sodium hydroxide (NaOH) ratio up to 2.5, the greater strength and durable geopolymer to be produced. This is because sodium silicate affecting the microstructure of geopolymer which leads to structural strength. However, higher of sodium silicate to sodium hydroxide ratio, the quantity of sodium hydroxide is not sufficient for the completion of dissolution process during the geopolymerization which could lead to decreasing in strength. Next, the higher the curing temperature up to 100°C, the greater strength and more durable geopolymer can be produced. More than that, it will cause shrinkage and cracks on the geopolymer which resulting in the loss of strength. Next, curing duration is the key in obtaining higher average strength value. When the curing duration is prolonged, the strength of the geopolymer will be greater as longer curing time improved the polymerisation process. However, the strength improvement became less noticeable when the duration surpassed 48 hours. Moreover, precursor proportion affects the mechanical strength of the eggshell powder and fly ash based geopolymer. Increasing in fly ash content, will produce higher mechanical and durability properties of the geopolymer. Lastly, based on the literatures that have been reviewed, eggshell powder and fly ash based geopolymer can be utilised in the pavement construction.

References

[1] Dhanalakshmi M, Dr Sowmya N J, and Dr Chandrashekar A, "A Comparative Study on Egg Shell Concrete with Partial Replacement of Cement by Fly Ash," *Int. J. Eng. Res.*, vol. V4, no. 05, pp. 1532–1538, 2015.

[2] W. F. Ho, H. C. Hsu, S. K. Hsu, C. W. Hung, and S. C. Wu, "Calcium phosphate bioceramics synthesized from eggshell powders through a solid state reaction," *Ceram. Int.*, vol. 39, no. 6, pp. 6467–6473, 2013.

[3] P. Shekhawat, G. Sharma, and R. M. Singh, "Strength behavior of alkaline activated eggshell powder and flyash geopolymer cured at ambient temperature," *Constr. Build. Mater.*, vol. 223, pp. 1112–1122, 2019.

[4] S. Kamaruddin, W. I. Goh, A. A. Jhatial, and M. T. Lakhiar, "Chemical and Fresh State Properties of Foamed Concrete Incorporating Palm Oil Fuel Ash and Eggshell Ash as Cement Replacement," *Int. J. Eng. Technol.*, vol. 7, no. 4.30, p. 350, 2018.

[5] P. Shekhawat, G. Sharma, and R. M. Singh, "Microstructural and morphological development of eggshell powder and flyash-based geopolymers," *Constr. Build. Mater.*, vol. 260, p. 119886, 2020.

[6] M. D. A. Thomas, "Optimizing the Use of Fly Ash in Concrete," *Portl. Cem. Assoc.*, p. 24, 2007.

[7] P. Shekhawat, G. Sharma, and R. M. Singh, "Potential Application of Heat Cured Eggshell Powder and Flyash-Based Geopolymer in Pavement Construction," *Int. J. Geosynth. Gr. Eng.*, vol. 6, no. 2, pp. 1–17, 2020.

[8] T. Bakharev, "Geopolymeric materials prepared using Class F fly ash and elevated temperature curing," *Cem. Concr. Res.*, vol. 35, no. 6, pp. 1224–1232, 2005.

[9] N. A. Lloyd and B. V. Rangan, "Geopolymer concrete with fly ash," *2nd Int. Conf. Sustain. Constr. Mater. Technol.*, no. January, pp. 1493–1504, 2010.

[10] K. T. Nguyen, N. Ahn, T. A. Le, and K. Lee, "Theoretical and experimental study on mechanical properties and flexural strength of fly ash-geopolymer concrete," *Constr. Build. Mater.*, vol. 106, pp. 65–77, 2016.
[11] A. Palomo, M. W. Grutzeck, and M. T. Blanco, "Alkali-activated fly ashes: A cement for the future," *Cem. Concr. Res.*, vol. 29, no. 8, pp. 1323–1329, 1999.
[12] G. S. Ryu, Y. B. Lee, K. T. Koh, and Y. S. Chung, "The mechanical properties of fly ash-based geopolymer concrete with alkaline activators," *Constr. Build. Mater.*, vol. 47, no. 2013, pp. 409–418, 2013.
[13] J. Davidovits, "Geopolymers," *J. Therm. Anal.*, vol. 37, no. 8, pp. 1633–1656, 1991.
[14] D. Hardjito and B. V Rangan, "Development and Properties of Low-calcium Fly Ash Based Geopolymer LOW-CALCIUM FLY ASH-BASED GEOPOLYMER CONCRETE By Faculty of Engineering Curtin University of Technology," *Aust. Univ. Technol. Perth*, no. January, p. 48, 2005.
[15] P. Shekhawat, G. Sharma, and R. M. Singh, "Durability analysis of eggshell powder–flyash geopolymer composite subjected to wetting–drying cycles," *J. Eng. Des. Technol.*, vol. 18, no. 6, pp. 2043–2060, 2020.
[16] N. Cristelo, S. Glendinning, L. Fernandes, and A. T. Pinto, "Effect of calcium content on soil stabilisation with alkaline activation," *Constr. Build. Mater.*, vol. 29, pp. 167–174, 2012.
[17] N. Cristelo, S. Glendinning, L. Fernandes, and A. T. Pinto, "Effects of alkaline-activated fly ash and Portland cement on soft soil stabilisation," *Acta Geotech.*, vol. 8, no. 4, pp. 395–405, 2013.
[18] J. Davidovits, "Geopolymer Cement a review," *Geopolymer Sci. Tech.*, no. 0, pp. 1–11, 2013.
[19] GRAY DH and LIN YK, "Engineering Properties of Compacted Fly Ash," no. March 2006, 1971.
[20] B. Joseph and G. Mathew, "Influence of aggregate content on the behavior of fly ash based geopolymer concrete," *Sci. Iran.*, vol. 19, no. 5, pp. 1188–1194, 2012.
[21] D. Hardjito and S. E. Wallah, "Effect of mixing time and rest period on the engineering properties of fly ash-based geopolymer concrete," *Aust. J. Struct. Eng.*, vol. 6, no. 1, 2005.
[22] P. P. Rachel and D. P. P. Dr. P. Partheeban, "Durability Study of Low Calcium Flyash Based Geopolymer Concrete," *Indian J. Appl. Res.*, vol. 3, no. 6, pp. 180–182, 2011.
[23] Y. Ding, J. Dai, and C. Shi, "Mechanical Properties of Alkali-Activated Concrete : A State-of-the-Art Review Keywords," 2016.
[24] M. F. M. Tahir, M. M. A. B. Abdullah, M. R. M. Hasan, and W. W. A. Zailani, "Optimization of fly ash based geopolymer mix design for rigid pavement application," *AIP Conf. Proc.*, vol. 2129, no. July, 2019.
[25] K. Sagoe-Crentsil and L. Weng, "Dissolution processes, hydrolysis and condensation reactions during geopolymer synthesis: Part II. High Si/Al ratio systems," *J. Mater. Sci.*, vol. 42, no. 9, pp. 3007–3014, 2007.
[26] L. Weng and K. Sagoe-Crentsil, "Dissolution processes, hydrolysis and condensation reactions during geopolymer synthesis: Part I-Low Si/Al ratio systems," *J. Mater. Sci.*, vol. 42, no. 9, pp. 2997–3006, 2007.
[27] P. Sukmak, P. De Silva, S. Horpibulsuk, and P. Chindaprasirt, "Sulfate Resistance of Clay-Portland Cement and Clay High-Calcium Fly Ash Geopolymer," *J. Mater. Civ. Eng.*, vol. 27, no. 5, p. 04014158, 2015.

Acknowledgments

Thanks are due to the Sdg-borneo (SDG-BORNEO) for funding the projects 600-RMC/SDG-BORNEO 5/3 (017/2020).

Landslide Runout Distance Prediction of Pinousuk Gravel Slope in Mesilou Kundasang, Sabah Using Slope Characterization

H F W S Erfen[1,2*] and B Musta[1,2]

[1] Faculty of Science and Natural Resources, Universiti Malaysia Sabah, Jalan UMS 88400, Kota Kinabalu, Sabah, Malaysia
[2] Natural Disaster Research Centre, Universiti Malaysia Sabah, Jalan UMS 88400, Kota Kinabalu, Sabah, Malaysia

*Corresponding author: babamus@ums.edu.my

Abstract. The study area is located at Mesilou, Kundasang, which focuses on slopes of the Late Pleistocene Pinousuk Gravel unit. This study aims to determine the landslide runout distance (L) against the slope height (H), slope angle (θ), percentage of sand and clay of slope materials. Data were obtained from a shaking table instrument used to simulate earthquake tremors that resulted in landslides. The results show that the higher the slope it is, the farther the runout will occur with correlation, R = 0.87. Based on the slope failure model, the increase of slope angle will add to 60% of the runout distance with R = 0.81, while the longer time of the tremor received increases more than 120% of the distance from the previous point. Results also show insignificant correlation between clay percentage with runout distance (R = 0.23) while higher percentage of sand shows a greater distance from the slope. The greater distance of movement was due to its incohesive and unconsolidated characteristics with correlation, R = 0.64. All the findings show that slope characteristics, such as slope geometry and slope soil properties, influence the distance of landslide runout.

Keywords: Landslide, runout prediction, shaking table, Pinousuk Gravel

1. Introduction

Landslide, a natural hazard, commonly occurred either is triggered by seismic activity or high rainfall reception. The impact of landslide can be observed through the destruction of properties and fatalities. Thus, the understanding of landslide's behaviour is vital for hazard mitigation and risk assessment mainly to the vulnerable locations. The prediction for landslide runout distance is essential as a safety measure for buffer zone construction to curb the impact of landslide. Several methods were introduced to predict the runout distance including using center of mass (COM) approach, 3D models and statistical analysis [1, 2, 3]. It also can be estimate that the runout distance is influenced with the characteristics of the slope and its materials. Therefore, the aim of this paper is to predict the runout distance of landslide which triggered by seismic simulation by taking consideration of slope geometry and properties of slope materials.

The study area is located at highland of Mesilou, Kundasang, Sabah which has experienced numerous impacts from Ranau Earthquake 2015 such as landslides and debris flows. The area consists of several rock units but focuses on Pinousuk Plateau which comprises of Pinousuk Gravel Unit (Figure 1). Pinousuk Gravel is a tilloid deposits from the glaciation of Mount Kinabalu during Late Pleistocene around 40000 years ago [4]. It is formed by two depositional phases and is characterized with unconsolidated, weak materials. The location is within Lobou-Lobou Fault and Mensaban Fault, makes the area prone for any landslide occurrences [5].

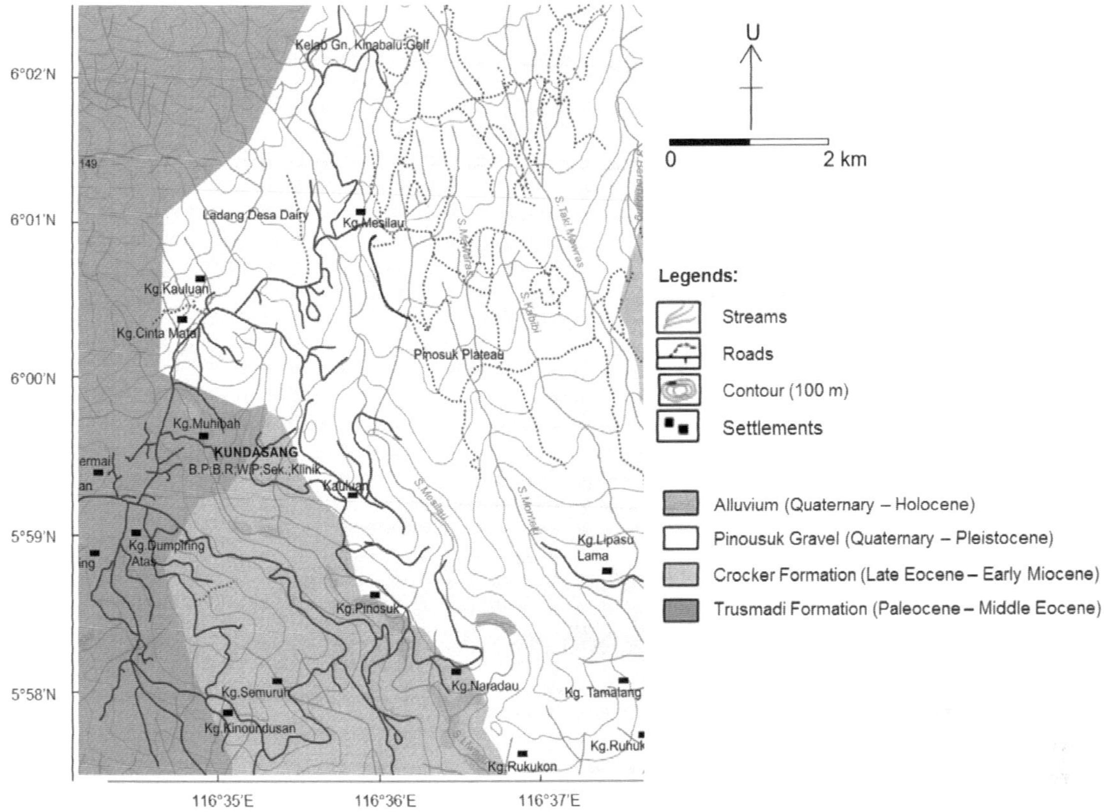

Figure 1. Geological map of the study area

2. Methodology

For this study, shaking table method is used where simulation of seismic tremor is formed to be applied on the slope model. Shaking table method has widely conducted to study the failure design of slope when experienced earthquake [6, 7, 8]. Time series of earthquake event on 7 March 2018 with 5.2 Richter scale in Ranau is selected to generate simulated waveform (Figure 2). It is later applied on the slope model which is designed in test box of 10 cm x 20 cm in height and length, respectively, using uniform sand and clay with 35° angle. The slope model is subjected to 15 seconds to 30 seconds vibration to observe the failure design and the runout distance (Figure 3). Similitude law is used to correlate the result of simulation runout with the actual slopes [9] using scale factor λ which is the height ratio between actual slope and the slope model. The runout distance is later recalculated again using actual percentage of sand and clay of research slopes to estimate the in-situ runout distance of the study area.

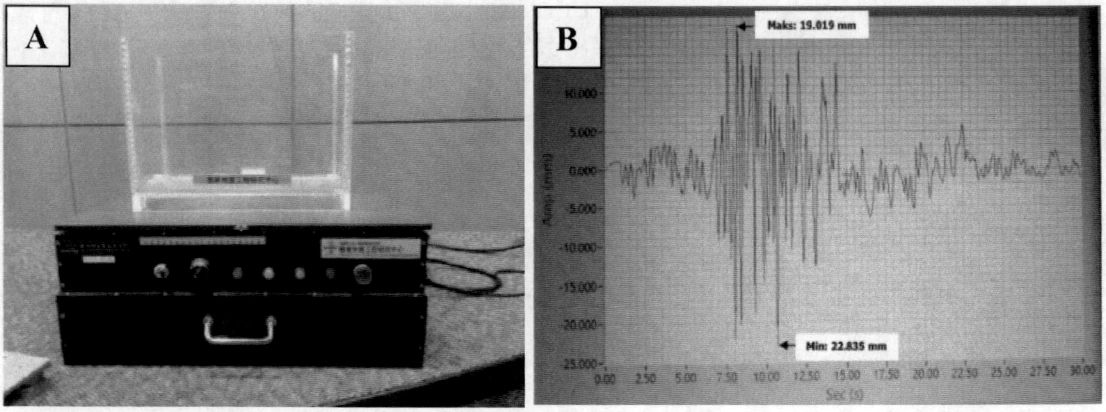

Figure 2. (A) Shaking table; (B) Simulated waveform from earthquake event in Ranau on 2018

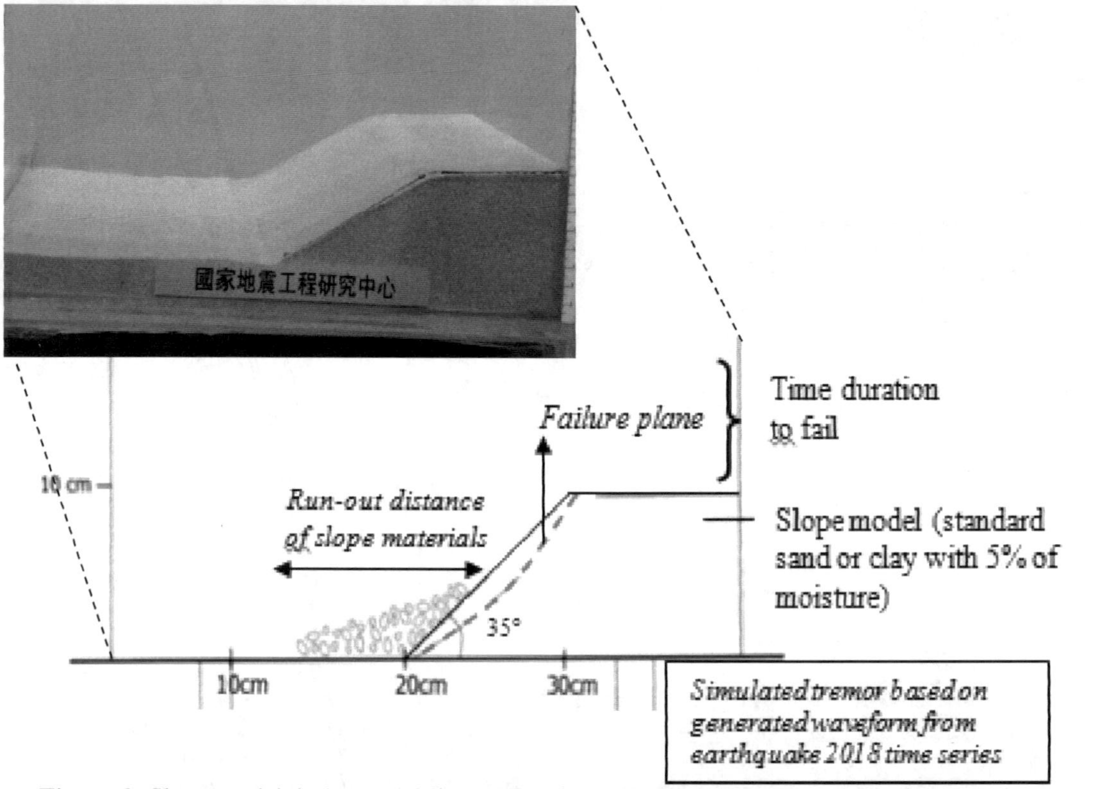

Figure 3. Slope model design and failure estimation (Small box: slope model before tested)

Four slopes from Pinousuk Gravel unit (labelled as S1 to S4) (Figure 4) were chosen with different lithology and locations. S1 is matrix dominated of ultramafic soils, S2 is dominated with ultramafic rocks clasts, S3 consists of smaller fragments of ultramafic rocks and active fault lines and S4 is dominated with granodiorite soil and boulders. All slopes are located nearby to the buildings and main road causing high risk if landslide occurs.

Figure 4. Research slopes in the study area from Pinousuk Gravel Unit of unconsolidated materials

3. Results and Discussion

Table 1 below shows the data for slope geometry and soil properties. These parameters are later used for interpreting the possible runout distance in the study area from simulation data which obtained from the shaking table method.

Table 1. Data for slope geometry and soil properties of research slopes

	S1	S2	S3	S4
Slope Height	7.11 m	6.32 m	11.80 m	3.95 m
Slope Angle	55°	70°	82°	71°
Moisture Content	61.74%	42.93%	53.49%	43.51%
Plasticity	30.48%	25.22%	16.19%	13.31%
Sand	28.91%	60.34%	34.39%	48.56%
Clay	26.10%	31.83%	29.16%	30.35%

Based on Erfen [10], the safety factors of research slopes range from 1.15 to 2.63 which indicate as stable slopes. However, the cohesion for soil slopes which acted as the binding material for the clasts are low (1.6 kN to 2.8 kN) [10]. It exhibits varies of moisture value and texture which reduce the ability to hold and facilitate for loose material to slide.

3.1 Runout Distance Prediction

From shaking table method, the slope model showed runout distance of 5 cm from the slope toe when experienced first tremor. It later getting 18 cm farther when subjected to continuous tremor which is twice or more than 120% than the first runout distance (Figure 5 and Figure 6). This is consistent with landslides from Wenchuan Earthquake in China where the movement of the material travel for short distance but continue to move away when receive stronger or continuous tremors [11]. It is also supported by Havenith and Bourdeau [12] that slope materials can move further on longer tremors. This can be seen from recalculation using percentage of sand and clay of actual slopes, the runout distances range from 2.5 meters to 6.0 meters during first vibration, before moving farther from 5.6 meters to 13.5 meters after second vibration (Table 2).

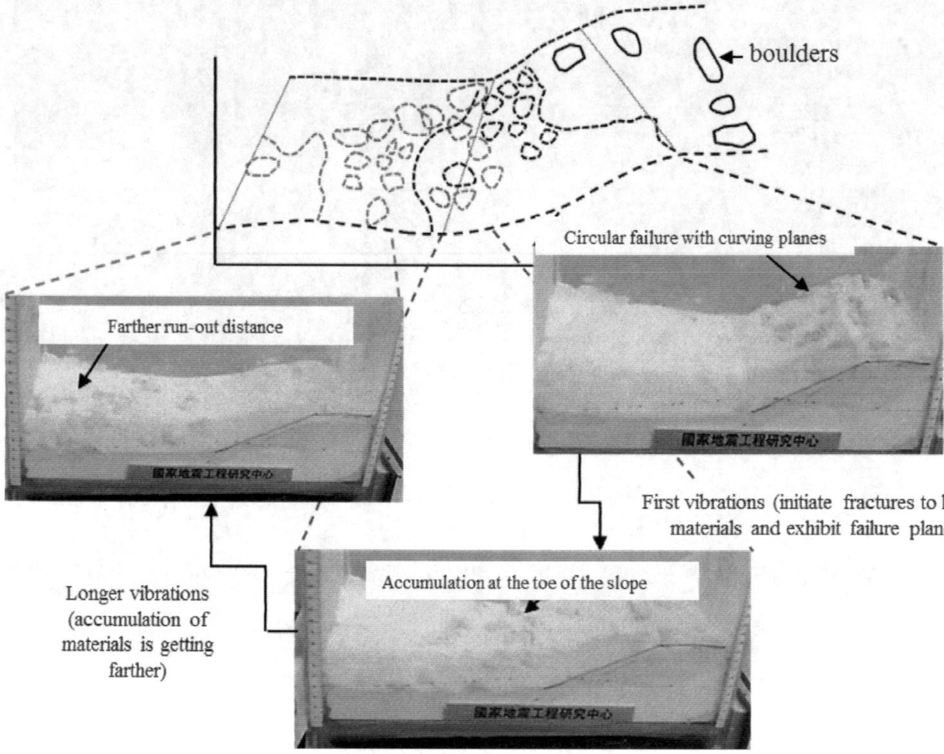

Figure 5. Descriptive sketches of run-out distance of slope material during first shaking (15 seconds) and continuous shaking received.

Table 2. Runout distance prediction using simulation failure from slope model

	S1	S2	S3	S4
Slope Model				
Runout During First Vibration (m)	0.05	0.05	0.05	0.05
Final Runout After Second Vibration (m)	0.18	0.18	0.18	0.18
Actual Slope (Recalculation)				
Slope Height (m)	7.11	6.32	11.80	3.95
% Sand	28.91	60.34	34.39	48.56
% Clay	26.10	31.83	29.16	30.35
(Total %)	*55.01*	*92.17*	*63.55*	*78.91*
Runout during First Vibration (m)	3.1	4.7	6.0	2.5
Final Runout After Second Vibration (m)	7.0	10.5	13.5	5.6

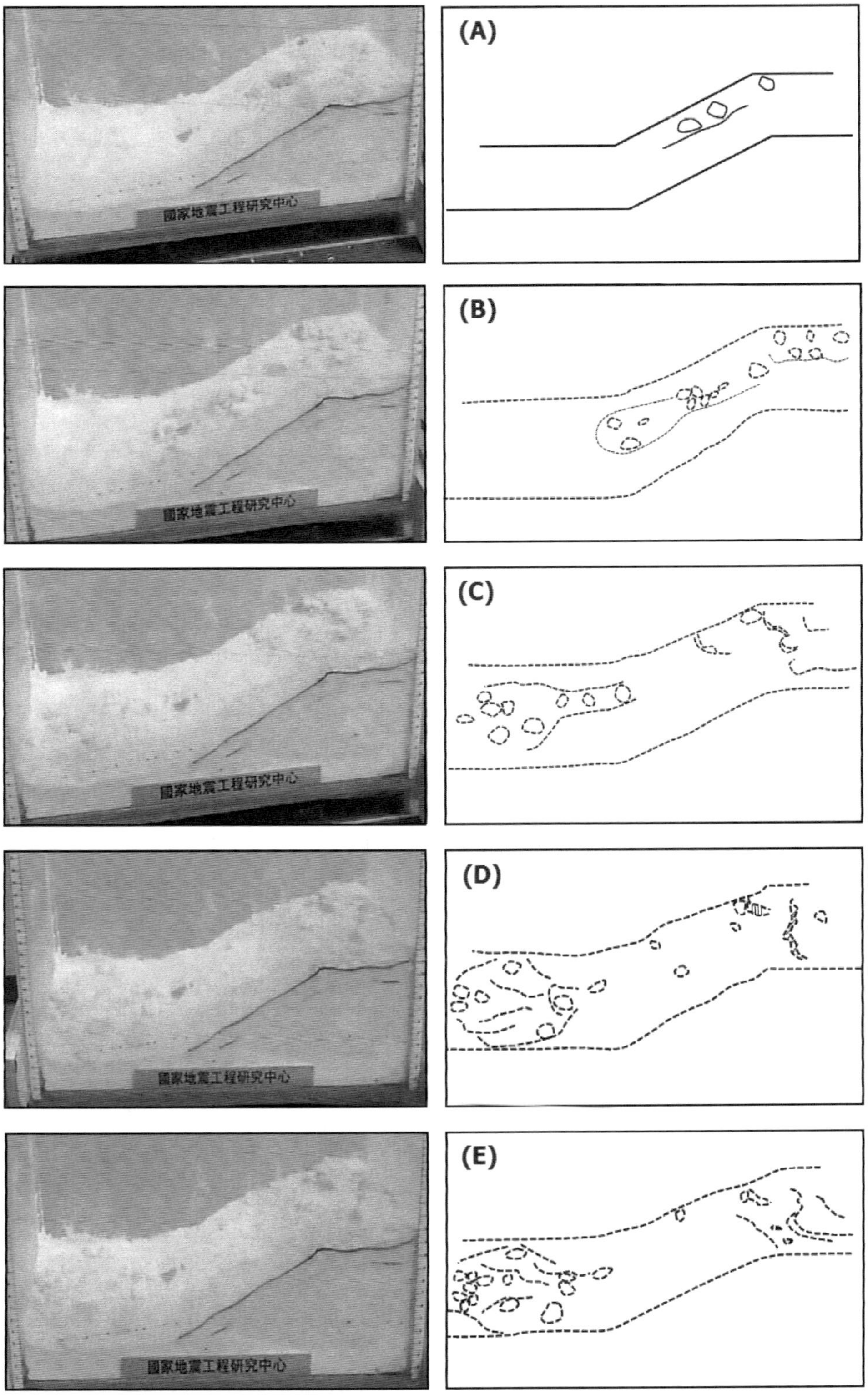

Figure 6. Diagram (A) to (C) showed the failure and movement of materials after first vibrations of 15 seconds, followed by (D) and (E) to exhibit farther run-out distance of debris accumulation of 18 cm from the toe of the slope model when received continuous vibration

3.2 Correlation with Slope Characterization and Soil Properties

The slope height also plays an important role in estimating the runout distance. The increase in slope height will influence the slope material to move further. Figure 7 show slope geometry of height and angle of slope which linearly correlated, with value of R is 0.87 and 0.81, respectively (Table 3). R value more than 0.8 show strongly correlated between two variables which indicate farther runout distance with the increase of slope height and angle.

Table 3. Correlation between slope geometry and soil properties with runout distance

	L
Runout distance, L	1
Slope height, H	0.870
Slope angle, tan (θ)	0.808
Moisture content	-0.561
Plasticity	-0.132
% Sand	0.643
% Clay	0.229

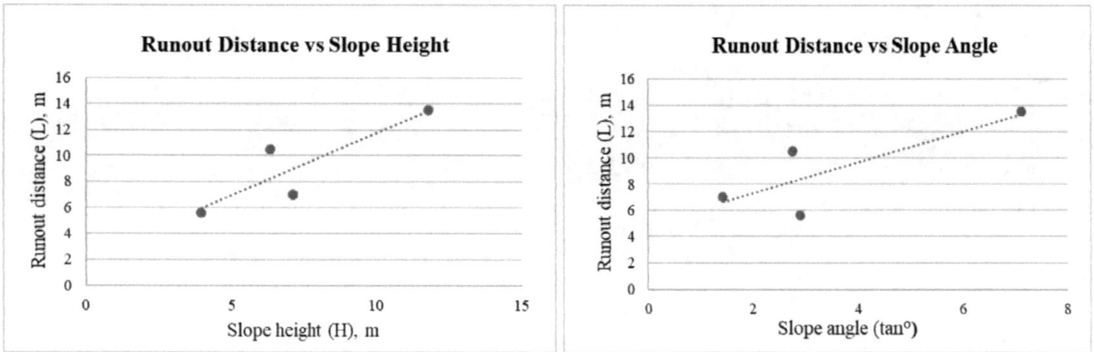

Figure 7. Strongly correlated between runout distance with the slope height and angle

Slope with higher sand percentage leads to larger runout distance. This is due to cohesionless materials with less moisture accelerates the materials to travel faster and farther [1,13]. The correlation between sand percentage and runout distance for slopes in the study area show moderate correlation with R = 0.64. However, clay percentage in the study area show small and negligible correlation to influence the runout distance (R = 0.23). High percentage of clay with high fraction of moisture may initially enhances the runout [14] but reduce it when too much clay forming viscous flow [13]. Figure 8 below shows the correlation between percentage of sand and clay with the runout distance. It exhibits moderate to insignificant correlation, relatively, which depending on other factors such as moisture content that might affects the distance of runout prediction.

Based on Pinousuk Gravel slopes in the study area, slope S1, S2 and S3 which consist of ultramafic materials have larger runout distance between 7 meters to 13.5 meters, than slope S4 which has granodiorite materials with only 5.6 meters runout. This shows that each parameter of slope geometry and soil texture must be considered together to predict better runout distance for any landslide occurrence in the future.

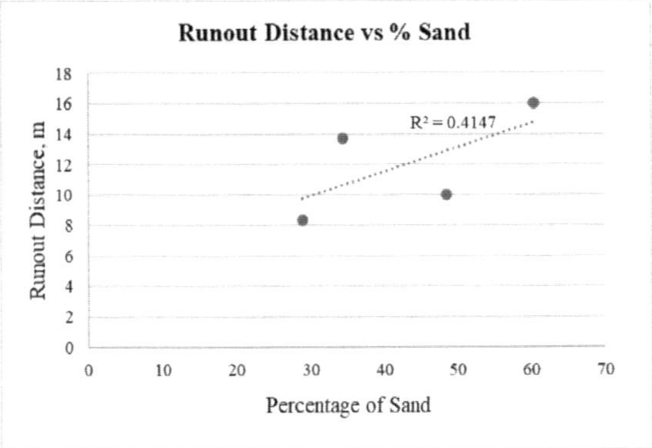

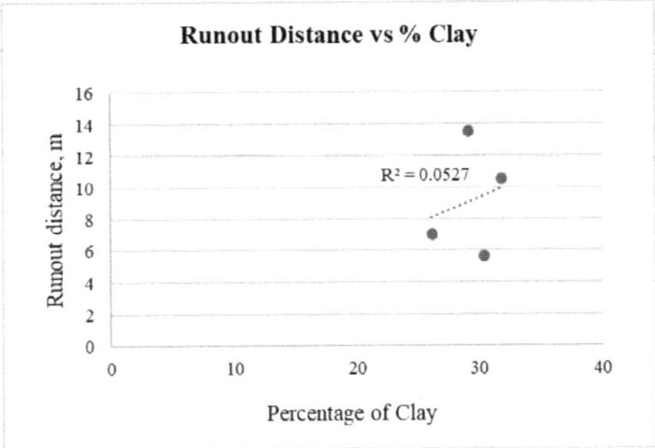

Figure 8. Moderately correlated between runout distance with percentage of sand (left); negligible correlation between runout distance with percentage of clay (right)

4. Conclusions

The research is focused on Pinousuk Gravel slopes to predict the runout distance using computational method from a simulation of earthquake-induced landslide model. Results showed that slope geometry (height and angle of slope) are strongly correlated with the runout distance with $R < 0.80$. The percentage of sand and clay also influenced the runout distance depending on the moisture content, plasticity and cohesion of the slope material with moderate and negligible correlation, respectively.

5. References

[1] Firmansyah E, Selly F, Adrin T and Fourier D E L 2016 *IOP Conference Series Earth and Environmental Science.* **29**(1) 1-6
[2] Hungr O 1995 *Can. Geotech. J.* **32** 610-623
[3] Devoli G, De Blasio F V, Alverhoi A and Hoeg K 2009 *Geotch Geol. Eng* **27** 23-42
[4] Marilah Saman and Ibrahim Komoo 2000 *Geological Society of Malaysia Annual Geological Conference*
[5] Azhari Mohamed 2012 *19th United Nations Regional Cartographic Conference for Asia and the Pacific*
[6] Wang K L and Lin M L 2011 *Engineering Geology* **122** 106-114
[7] Song D, Che A, Zhen Z and Ge X 2019 *Engineering Geology* **245** 153-168
[8] Zhao Z, Ren C, Xu G, Zhan H and Liu T 2019 *Shock and Vibration* 1-19
[9] Lim M L and Wang K L 2006 *Engineering Geology* **86**(2) 118-133
[10] Erfen H F W S, Musta B and Ahmad M S 2019 *Journal of Physics: Conference Series* **1358** 1-11 012070
[11] Wang F, Sun P, Highland L and Cheng Q 2014 *Geoenivronmental Disasters* **1**
[12] Havenith H B and Bourdeau C 2010 *Geologica Belgica* **12**(3) 137-152
[13] De Haas T, Braat L, Leuven J R F W, Lokhorst I R and Kleinhans M G 2015 *J of Geophysical Research: Earth Surface* **120**(9) 1949-1972
[14] Hungr O, Evans S G, Bovis M and Hutchinson J N 2001 *Environmental and Engineering Geoscience* **7**(3) 221-238

Acknowledgments

The author would like to express heartiest appreciation to Universiti Malaysia Sabah for funded this research under Research Grant code SBK0276-STWN-2016 and Special Grant code SDK0015-2017. Special thanks to Dr. Chang Yen-Wei and Dr. Kuo Chun-Hsiang from National Dong Hwa University and National Centre of Research for Earthquake Engineering, Taiwan for permission and assistance while handling the shaking table instrument.

The efficiency of sewage treatment plant: A case study at the main campus of Universiti Malaysia Sabah (UMS)

N. Bolong[1,2]*, M. Q. N. Sabli[1], I. Saad[1,] and A. N. A. Ali[1,3]

[1] Faculty of Engineering, Universiti Malaysia Sabah, 88400 Kota Kinabalu, Sabah, Malaysia.
[2] Green Materials and Advanced Construction Technology (GMACT) Research Unit, Universiti Malaysia Sabah, Malaysia.
[3] Faculty of Civil Engineering, Universiti Teknologi MARA, Cawangan Sarawak, 94300 Kota Samarahan, Malaysia.

*Corresponding author: nurmin@ums.edu.my https://orcid.org/0000-0002-4035-7173

Abstract. The aquatic nature, including streams, rivers, marine, and groundwater, becomes susceptible to most contaminants' effects through various pathways known as point and non-point sources. Point source pollution from inefficient sewage treatment plants (STP) will likely produce effluents that could deteriorate the environment. Therefore, this study assesses the efficiency of STP at Universiti Malaysia Sabah (UMS) based on the Malaysian Sewerage Industry Guidelines (MSIG). The design parameters of the STP and its influent and effluent physicochemical parameters are evaluated and compared between usage peak time (lecture-week) and non-peak (semester-break) conditions. The measurement was based on the Standard Methods of Examination of Water and Wastewater. The permissible limits of influent and effluent from Malaysia Sewerage Industry Guidelines (MSIG) are referred. In contrast, the efficiency of the STP is calculated as the percentage reduction of pollutants. The findings revealed that the current STP's design parameters complied with the MSIG standard. The suspended solids (TSS), biochemical oxygen demand (BOD), chemical oxygen demand (COD), oil and grease (O&G) at influent during peak time is typically higher than during the non-peak time, albeit a timely sudden increase can be observed. The effluent quality is inconsistent but is still within permissible limits, except for O&G, which occasionally exceeds the allowable value. As for the BOD, a consistent 80% removal efficiency is observed. Hence the effluent reuse for agricultural purposed and landscape irrigation is proposed due to the environmental quality potential and reduction to direct released toward the receiving water streams.

Keywords: Malaysian Sewerage Industry Guidelines (MSIG), physicochemical parameters, removal efficiency, sewage treatment plants (STP), wastewater

1. Introduction

The Sewage treatment plants (STP) undeniably have a vital role in maintaining public health in an urbanized community. A typical septic tank would be sufficient for a small house in a rural area, but not in large populations such as shopping complexes, residential areas, schools, and universities. Unlike a water treatment plant (WTP) that primarily treats water for drinking purposes, the STP ensures the treated wastewater is safe when discharged to the environment.

Processes for sewage treatment can be categorized into physical, chemical, and biological. Nonetheless, the treated effluent may not be potable due to ions, metals, and microorganisms that could harm humans [1-3]. To help plan, design, and construct STP, Malaysia Sewerage Industry Guidelines (MSIG) Vol. 4 (2010), which was derived from Environmental Quality Act 1974 (EQA 1974), has been published by National Water Service Commission (NWSC) for local use. This standard also presents the permissible limits on effluent discharges that emphasize various physicochemical parameters.

Physicochemical parameters refer to both the physical and chemical properties of the wastewater [4]. Examples of physical properties are temperature, pH, turbidity, total dissolved solids (TDS), and TSS. As for chemical parameters, it includes dissolved oxygen (DO), biochemical oxygen demand (BOD), chemical oxygen demand (COD), nitrates, and total phosphate. The efficiency of STP can be determined by calculating the percentage of reduction from these selected parameters. Though BOD, COD, and TSS are the most commonly used to investigate the efficiency of STP [5-7].

Wastewater can be detrimental to the environmental ecosystem and even affect human health without appropriate management [8]. Thus, the design and operation of STP are critically vital to ensure the wastewater can be safely released back into nature without adverse impact. However, building STP is expensive [9]. Therefore, design requirements based on the permissible limits need to be determined to ensure that the most suitable treatment process is used, optimizing the installation and operational costs. The appropriate design includes the technology applied, types of media used, and cost for land acquisition.

In dealing with various treatment technology, it is essential to consider factors such as the size of the area needed to construct, the treatment capacity, and its by-products [10]. More capital can be saved when only a small space is required for the plant, and costs can be reduced when treatment can be done efficiently in a large quantity of wastewater with fewer sludge by-products to be dealt with. For instance, the use of sequencing batch reactor (SBR) and membrane bioreactor (MBR) during wastewater treatment produces removal efficiency of more than 90% for BOD and COD [11]. Compact yet highly efficient treatment is also more favorable when limited space, such as commercial buildings or limited land building areas. There are also options of using other techniques, such as utilizing microalgal biofilms in post-treatment [12] or having multi-layered filters [13].

Transpiring the concept of converting waste to wealth, using recyclable materials such as construction wastes and agro-wastes as filter media to improve effluent quality has also attracted the industry [14, 15]. Utilizing these materials has shown great potential in reducing even metals and microorganisms [16, 17]. However, a continuous study is still ongoing worldwide due to the complexities of variables involved when processing the materials.

Another possible focus is on reusing the wastewater for non-potable uses as an alternative to the existing water source. The limits of the local effluents are bound under the Environmental Quality (Sewage) Regulations 2009 and Environmental Quality (Industrial Effluent) Regulations 2009. For example, only effluent with BOD of less than 20 mg/L is permissible to be discharged, according to Standard A's requirements. Hence, it is vital to identify the main mechanisms involved in removing the desired contaminants [18] and optimize the systems to fit the treatment needs [19] without compromising the receiving water source. Nonetheless, the perception of reusing wastewater requires more attention, and awareness needs to be promoted that it is safe to be used since most pollutants have been removed according to the required guidelines.

2. Methodology

The study area of Universiti Malaysia Sabah (UMS) main campus located in Kota Kinabalu, Malaysia. With approximately 999 acres, this university currently enrolls more than 13K students from various faculties. The location of UMS Central-STP (6.038187, 116.130071) is shown in Figure 1.

The main objective of this paper is to evaluate the Sewerage Treatment Plant (STP) of the UMS institution. Figure 2 shows the flow of methodology conducted.

Figure 1. Location of UMS Central-STP.

2.1. *Consultation with the Department of Development and Maintenance Office (DDM)*

This study incorporated the technical drawing of STP, operational and maintenance practices, and existing effluent monitoring procedures with permission from the Department of Development and Maintenance (DDM), which operates and manages the STP. First, the initial design of the STP was obtained from the DDM. Then the results of water quality testing from the appointed consultant were compared with the current study.

2.2. *Water Sampling and Testing*

Grab sampling was used to collect water samples as recommended in Malaysia Sewerage Industry Guidelines (MSIG) Volume 4 [20]. The sampling was conducted for six days, 3-days during the non-peak time (semester-break), and another 3-days during the peak time (lecture-week). On the collection day, water samples at both influent and effluent were collected during the morning (9:00 am to 10:00 am) and evening (2:00 pm to 3:00 pm) to study variations of physicochemical parameters in a single day. Respective sample collections and preservation follows (APHA 1060B) Collection of samples; whereas Table 1 summarizes the sampling collection time with the measured parameters and its standard method based on the Standard Method for Examination of Water and Wastewater [21]. The Spectrophotometer HACH DR6000 was used to obtain the COD and TSS. All results and secondary data from the DDM operator were compared with the Malaysia Sewerage Industry Guidelines (MSIG) permissible limits.

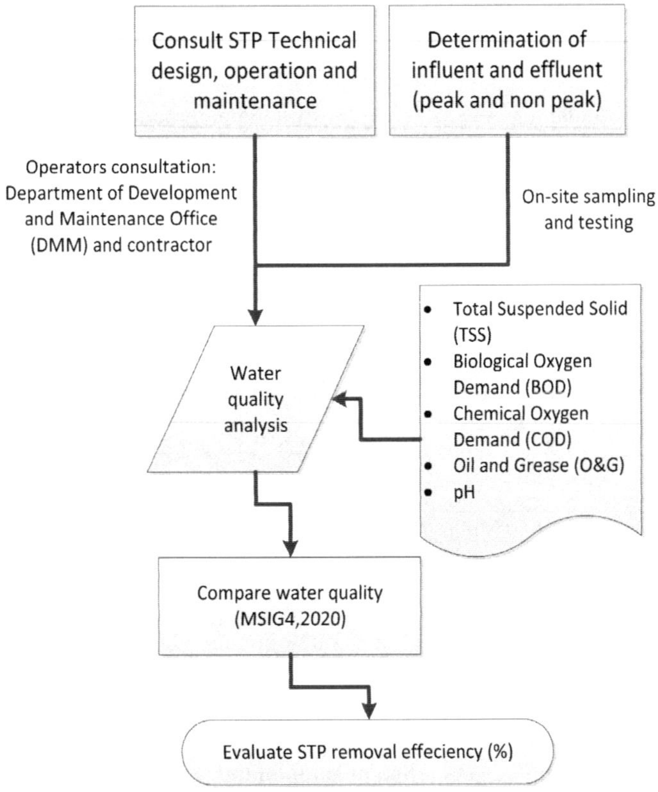

Figure 2. Methodology flow of this work

Table 1. The sampling collection and selected parameters with standard references.

Sampling time	Measured parameters	Standard method
Non-peak time *(Semester-break): 17, 18, 21 January 2019*	pH	(APHA 4500-H+B) Electrometric method
	Total Suspended Solid (TSS)	(APHA 2540A(3)) Sample handling and preservation
Peak time *(Semester week): 15 February 2019, 8, 26 March 2019*	Biochemical Oxygen Demand (BOD)	(APHA 5210B) 5-day BOD test
	Chemical Oxygen Demand (COD)	(APHA 5220D) Closed reflux, Colorimetric method
	Oil and Grease (O&G)	(APHA 5520B) Partition-Gravimetric method

2.3. Evaluation on the Removal Efficiency of STP

The STP efficiency (%) is calculated using equation (Ci-Ce)/Ci x 100, where Ci and Ce are the influent and effluent concentrations of selected parameters (TSS, BOD, COD, and O&G), respectively. All concentration values recorded are in milligram per liter (mg/L). The more significant gap between the initial and final concentrations, the higher the efficiency of the STP. Reduction of pollutants that achieved at least 80% of removal is already considered efficient regardless of the types of technology or technique being implemented, as suggested by previous studies [6, 22]. The measured TSS, BOD, COD, and O&G may be limited as other unregulated or not measured pollutants may also be harmful to the environment. Because some revolutionized and emerging contaminants contributed from sewage treatment as most current or conventional STPs are not designed to treat the persistent compounds and metabolites [23].

3. Results and Discussion

3.1. Technical design and operation of UMS-Central STP

Typically, population equivalent (PE), flow range, influent sewage, and performance are part of the factors considered when designing for treatment [20, 24]. Table 2 tabulate the design parameters of the UMS-Central STP. This STP was designed for 20,000 Population Equivalent (PE) (Class 3) to cater to 4,500 meters cubic per day (m^3/day). Here, a sequencing batch reactor (SBR) is used, which primary function is to treat the BOD and COD, compared to activated sludge treatment (AST) [11]. In addition, since the primary treatment uses a single tank, less area space is needed for the design [6, 25]. Indirectly this reduces the capital costs for the technology implementation.

There are two SBR tank units in this STP, where each tank can accommodate half of the actual design parameters. This design complies with the modularization requirement from the standard, where a minimum of two module units is required for STP of Class 3. Employing two units allows the STP to cater to a high workload, particularly during peak flow [26]. In addition, this ensures that the treatment process is not affected if maintenance is needed as continuous operation can be done while the other unit is maintained.

The effluent permissible limits from Standard A is more stringent than Standard B. Thus, this study uses Standard A for comparison purposes. Worth noting that Standard A is used when the treated effluent discharge point is upstream from a water intake point. In contrast, Standard B is applied when the treated effluent discharge point is downstream from a water intake point for consumption or water catchment areas. The schematic UMS STP design is simplified in Figure 3.

Table 2. Design parameters of UMS-Central STP [27].

Design parameter	
Design population equivalent	20 000 PE
Design average flow	4500 m^3/day
Number of SBR tank unit	2
For each respective SBR tank unit	
Design population equivalent	10 000 PE
Design average flow	2250 m^3/day
Influent BOD concentration	251 mg/L
Influent suspended solids	303 mg/L
Required effluent	Standard A
Effluent BOD	20 mg/L
Effluent suspended solids	50 mg/L
Aerobic solids retention time	23 days
Food: Microorganism ratio	0.08

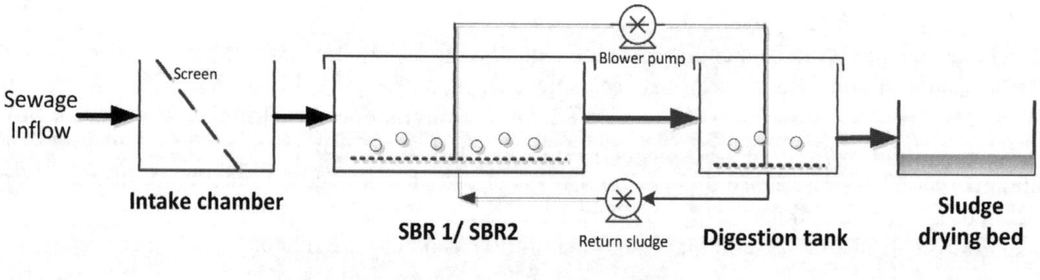

Figure 3 The schematic of the UMS STP system

3.2. Effect of sampling peak time on the at Influent and Effluent of STP parameters

It is usual for influent to have a nearing neutral pH value since raw wastewater typically ranges from 5.5 to 8.0. However, the observed pH obtained in this study ranges from 5.6 to 6.7, which is within the expected range. As for the effluent point, the pH does not show any significant changes, ranging from 5.8 to 6.9. The minor pH changes after treatment indicate that the alkalinity present in the wastewater is high [28]. The presence of alkalinity in wastewater provides a solid buffering capability that minimizes pH variation.

The measured trend of TSS, COD, BOD and O&G influent are illustrated in Figure 4 (a-b). It is observed that the trend of these measured parameters concentration at the influent point is generally lower during the non-peak time (semester-break). The concentration was higher in the evening as more sewage entered the STP, especially during peak time. The low TSS in the influent during the evening might be due to no rainfall event [29] as recorded during the sampling days.

At the influent sampling point, the measured COD was lower during the non-peak time (semester-break), slightly increased during the peak time (lecture-week), and higher in the evening. Thus, although they do not possess a direct relationship, COD usually is more elevated than BOD [30]. This indicates the amount of oxygen required to decompose organic matter and inorganic chemicals.

As expected, the BOD concentration at the influent is higher during peak time (lecture-week) and increased in the evening. Higher BOD potentially indicates that more biodegradable organics such as fats from the eateries, including cafeterias and food courts, leaked some of their waste to the drains.

Similarly, the measured O&G is slightly higher during peak time than during the non-peak (sem-break) period. However, all these influent values are still far less than design influent values as recommended by [20] and thus appropriately operational.

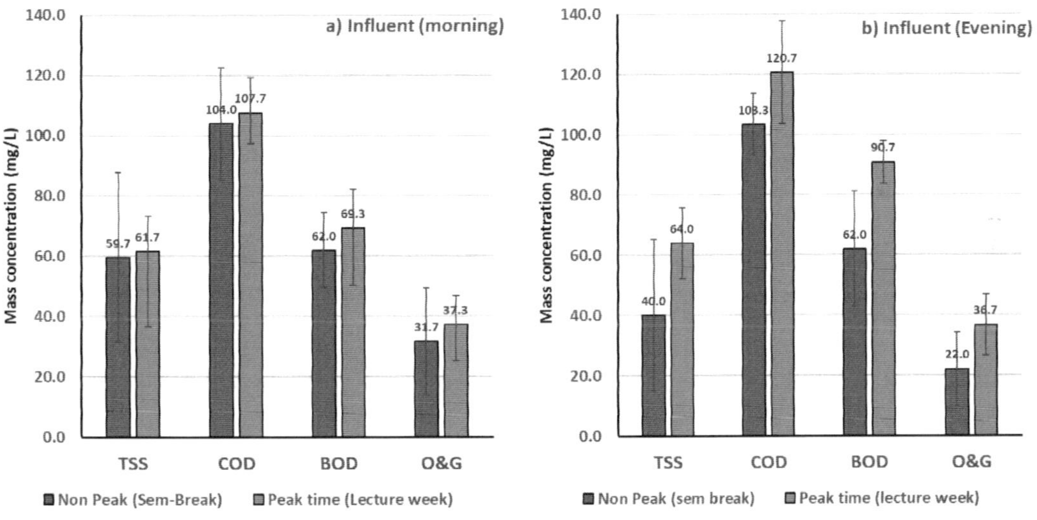

Figure 4 Variations of physicochemical parameters at the influent during non-peak and peak time sampling a) in the morning and b) in the evening

All parameters show effluent compliance with the permissible limits, as shown in Figure 5 a)morning and b)evening sampling. In addition, the effluent concentration is relatively consistent despite peak and non-peak times. Thus, the recorded parameters suggest that the UMS-Central STP consistently treated wastewater regardless of the concentration of the contaminants in the influent. A previous study reported that SBR is one treatment technology that can produce reliable effluent quality [11, 25].

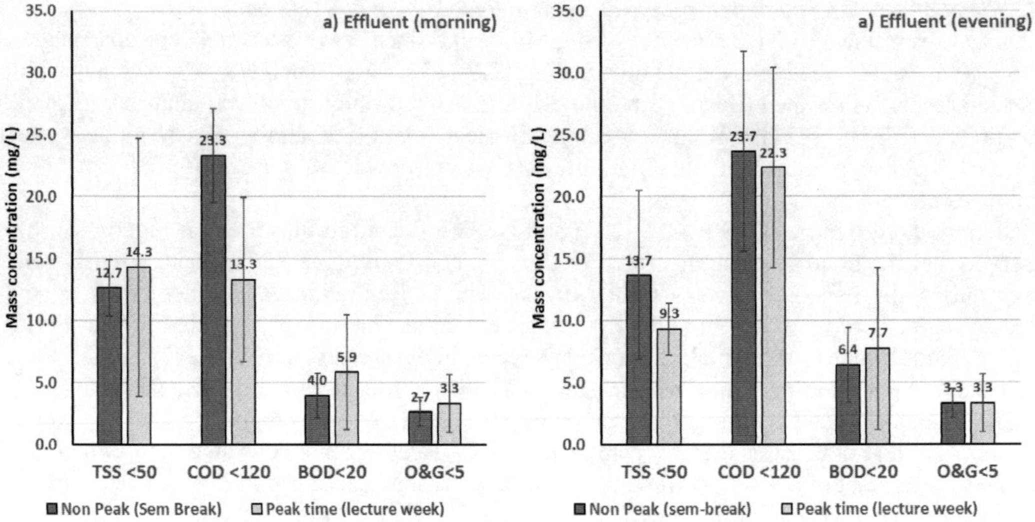

Figure 5. Variations of physicochemical parameters at the effluent sampling point during non-peak and peak time sampling a) in the morning and b) in the evening.

3.3. The efficiency of UMS-Central STP

Figure 6 (a-b) shows TSS, COD, BOD, and O&G removal efficiency for morning and evening, respectively. In comparing non-peak and peak time, the removal efficiency seems small (<5%) due to the consistent inflow. Furthermore, the STP has appropriately treated the sewage required by the standard as discussed previously. Consequently, the removal efficiency calculated based on the six-day sampling data is averaged and analyzed in Figure 6.

As in Figure 7, the removal efficiency for the measured sampling days found a slight decrease in removal efficiency during the evening. Based on the observed data, the average removal efficiency was slightly reduced in the evening, possibly attributed to higher concentration inflow during the daytime. However, BOD has the highest and most consistent removal efficiency of more than 80%. This result is also concurrent with previous findings, where BOD removal is very high regardless of influent variation in quality and quantity [6, 22]. TSS, however, has the lowest removal efficiency compared to other parameters. This could be due to the variation during operation when shifting between phases mainly to reduce the operational cost [31], increasing the TSS concentration in treated wastewater.

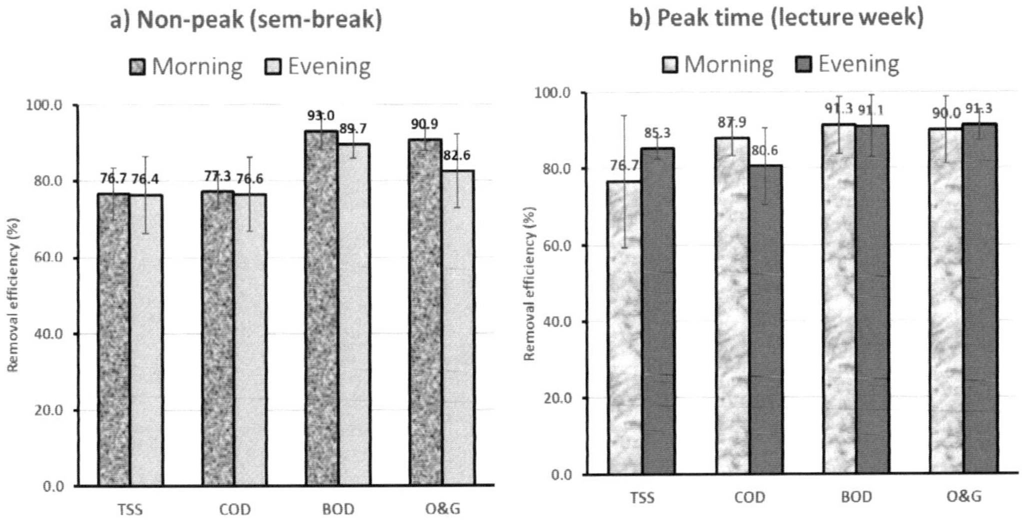

Figure 6. The removal efficiency of pollutants between a) non-peak and b)peak time

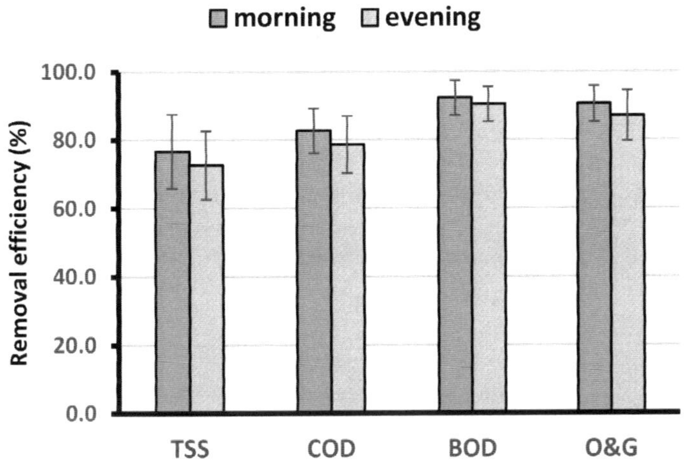

Figure 7. The average removal efficiency of STP.

3.4. Simple Cost-Benefit Analysis for Wastewater Reuse

Effluent water quality data for a three consecutive months duration that has been analyzed by the appointed consultant, as shown in Table 3, reveals that the UMS-Central STP complies with the standard used. Furthermore, all parameters were found to produce effluents of not more than their respective permissible limits. Except for recently, the TSS has exceeded the permitted level, indicating the maintenance need. Nonetheless, it is worth noting that variations of effluent can be expected due to the irregular natural flows and loads into the plants. Therefore, continuous sampling is vital to cross-checking the STP's efficiency.

Table 3. Results summary of water quality testing by the maintenance contractor and this work

Data source	Date	Effluent Parameters (mg/L)				Remarks
		TSS	COD	BOD	O&G	
This work	Jan 2019	14	25	3	2	Non-peak (sem break)
	Feb 2019	6	10	5	2	Peak time (lecture week)
	Mar 2019	26	21	11	2	Peak time (lecture week)
Maintenance Contractor	Jan 2019	16	47	7	3	Non-peak (sem break)
	Feb 2019	17	5	1	0	Peak time (lecture week)
	Mar 2019	11	11	5	0	Peak time (lecture week)
Maintenance Contractor	Jan 2021	7	42	2	1	Non-peak (Sem break)
	Feb 2021	52	2	2	1	Peak time (lecture week)
	April 2021	1	4	2	1	Peak time (lecture week)
Permissible limit (mg/L) [20]		<50	<120	<20	<5	Standard A (EQA, 1974, 2016)

While the effluents observed are not more than their respective permissible limits of Standard A [20], the released effluent to the receiving water is within the class IV (status suitable for irrigation) of WQI classification [32], regardless of dilution and concentration factor. Thus the potential for wastewater reuse is significant. From previous studies, the recommendation for wastewater reuse should be sufficient once treated and not used for irrigation purposes [33]. Understandably, reusing treated sewage effluent may have the unavoidable aesthetical reasoning for not being widely implemented. However, there have been many successful examples of wastewater recycling and reuse for non-potable applications, including landscape irrigation and agriculture, particularly for more than 90% removal efficiency [22]. More advanced treatment technologies are recommended to achieve this zero wastewater pollutants. This section discusses the cost-benefit that can be considered when reclaiming wastewater. Table 4 summarizes the cost-benefit qualitative comparison.

Table 4. The cost-benefit qualitative comparison for reclaiming wastewater.

Cost	Benefit
• Treatment and post-Treatment ↓	• Irrigation applications
• Monitoring and control ↑	• Abundance (mass-flow balance)
• Strict regulation and enforcement ↑	• Environment quality

Generally, there are two types of costs involved in wastewater treatment: fixed costs (for civil, mechanical, and electrical works) and variable costs (maintenance, labor, energy, chemicals) [34]. They also depend on the plant's capacity, with the target contaminants to be removed. In addition, the depreciation in the plant's lifespan may incur long-term planning for increasing the number of sub-plants. Nevertheless, consistent monitoring and control work is required to ensure the performance and lifespan of the plants can be maintained. Furthermore, optional technology such as integrating the Internet of Things (IoT) may also be considered [35] to aid in the monitoring and better enforcement adherent to the stringent environment guideline requirements.

In terms of the benefit, wastewater reuse is typically implemented for agricultural purposes and landscape irrigation [36], such as watering gardens and fields, and even commercial buildings (cleaning, toilet, flushing). Reclaim treated wastewater would increase crop production while reducing the seasonal yield variations since the wastewater is continuously available in abundance. The treatment of wastewater before reuse would also ensure safe discharge of the effluent into the environment according to standard, protecting both surface water and groundwater [37] and, eventually, nature. Thus, promote economic value while preserving the environmental ecosystem.

4. Conclusion and further works

Several conclusions made from the research work findings are:

- The primary design parameter of UMS-Central STP complies with the design requirements as stipulated in the referred (MSIG) Vol. 4 (2010). The current practices in the UMS STP adhered to the standard local requirement. Options on extending the STP effluent for reclamation purposes are recommended in this study to optimize sustainability and cost-benefit.
- The physicochemical parameters were successfully investigated and ranged from 75 to 90% removal efficiency. The results show that the treatment used can remove contaminants consistently, despite the initial contaminants' concentration in the influent.
- Overall, the average values of the measured STP effluent parameter lie approximately at the permissible limits, especially with O&G, whereas TSS, BOD, and COD parameters are within the allowable limits. Nonetheless, optimization during the treatment operation is vital to ensure the best possible output in treatment capacity, effluent quality, and costs.
- Wastewater is a precious resource when appropriately treated but occasionally overlooked. Future studies may continue to develop techniques and technologies that may materialize and promote wastewater more as an alternative to the water supply. In addition, a more detailed cost-benefit analysis would be beneficial in developing sewage treatments plants to prevent polluting water sources through advanced treatment yet natural methods.

References

[1] Leung, W. C., Wong, M.-F., Chua, H., Lo, W., Yu, P. H. F., & Leung, C. K. (2000). Removal and recovery of heavy metals by bacteria isolated from activated sludge treating industrial effluents and municipal wastewater. *Water Science and Technology*, *41*(12), 233–240. https://doi.org/10.2166/wst.2000.0277

[2] Gautam, S. K., Sharma, D., Tripathi, J. K., Ahirwar, S., & Singh, S. K. (2013). A study of the effectiveness of sewage treatment plants in Delhi region. *Applied Water Science*, *3*(1), 57–65. https://doi.org/10.1007/s13201-012-0059-9

[3] Tran, N. H., Reinhard, M., & Gin, K. Y.-H. (2018). Occurrence and fate of emerging contaminants in municipal wastewater treatment plants from different geographical regions-a review. *Water Research*, *133*, 182–207. https://doi.org/10.1016/j.watres.2017.12.029

[4] Nithyanandam, R., Huan, T. W., & Thao Thy, N. H. (2015). Case study: Analysis of water quality in Sungai Batu Ferringhi. *Journal of Engineering Science and Technology*, *10*(Special Issue), 15–25.

[5] Satyanarayana SV, S., Joodi, N., Saleh, A., & Kendy, A. (2015). Treatment efficiency study of Nizwa new sewage treatment plant. *International Journal of Applied Research*, *1*(8), 635–637.

[6] Aziz, H. A., Puat, N. N. A., Alazaiza, M. Y. D, & Hung, Y.-T. (2018). Poultry slaughterhouse wastewater treatment using submerged fibers in an attached growth sequential batch reactor. *International Journal of Environmental Research and Public Health*, *15*(8), 1734. https://doi.org/10.3390/ijerph15081734

[7] Kulkarni, B., Wanjule, R. V., & Shinde, H. H. (2018). Study on sewage quality from sewage treatment plant at Vashi, Navi Mumbai. *Materials Today: Proceedings*, *5*(1), 1859–1863. https://doi.org/10.1016/j.matpr.2017.11.286

[8] Chun, T. S., Malek, M. A., & Ismail, A. R. (2017). A review of wastewater treatment plant modelling: Revolution on modelling technology. *American Journal of Environmental and Resource Economics*, *2*(1), 22–26. https://doi.org/10.11648/j.ajere.20170201.13

[9] Chen, X., Xu, Z., Yao, L., & Ma, N. (2018). Processing technology selection for municipal sewage treatment based on a multi-objective decision model under uncertainty. *International*

Journal of Environmental Research and Public Health, *15*(3), 448. https://doi.org/10.3390/ijerph15030448

[10] Kawan, J. Ali, Hasan, H. A., Suja, F., Jaafar, O., & Abd-Rahman, R. (2016). A review on sewage treatment and polishing using moving bed bioreactor (MBBR). Journal of Engineering Science and Technology, 11(8), 1098–1120.

[11] Mirbagheri, S. A., SM, S. Y., Rafeidehkordi, N., & Vakilian, R. (2017). Comparison of CAS, MBR, SBR, and Biolak treatment systems in removal of BOD and COD from municipal wastewater - Case study: Ekbatan wastewater treatment plant. *15th International Conference on Environmental Science and Technology, Rhodes, Greece*, *31*(September), 1–5.

[12] Boelee, N. C., Temmink, H., Janssen, M., Buisman, C. J. N., & Wijffels, R. H. (2011). Nitrogen and phosphorus removal from municipal wastewater effluent using microalgal biofilms. *Water Research*, *45*(18), 5925–5933. https://doi.org/10.1016/j.watres.2011.08.044

[13] Onur, A., Ng, A., Batchelor, W., & Garnier, G. (2018). Multi-layer filters: Adsorption and filtration mechanisms for improved separation. *Frontiers in Chemistry*, *6*(417), 1–11. https://doi.org/10.3389/fchem.2018.00417

[14] Rahman, M. A., Imteaz, M. A., Arulrajah, A., Piratheepan, J., & Disfani, M. M. (2015). Recycled construction and demolition materials in permeable pavement systems: Geotechnical and hydraulic characteristics. *Journal of Cleaner Production*, *90*, 183–194. https://doi.org/10.1016/j.jclepro.2014.11.042

[15] A. Nasrizal, A. A., Bolong, N., & A. T., N. (2021). A review on the application of granular filter media and the utilization of agro-industrial wastes for stormwater quality improvement. *Jurnal Teknologi*, *83*(4), 1–16. https://doi.org/https://doi.org/10.11113/jurnalteknologi.v83.15159

[16] Ratnoji, S. S., & Singh, N. (2014). A study of coconut shell-activated carbon for filtration and its comparison with sand filtration. *International Journal of Renewable Energy and Environmental Engineering*, *2*(3), 201–204.

[17] Habib Abdullah, S. M., Niharul Haque, M., & Alamgir, M. (2015). Performance study of filler materials used in anaerobic baffled reactor. *Global Journal of Human-Social Science*, *15*(2), 24–30.

[18] Carballa, M., Omil, F., & Lema, J. M. (2007). Calculation methods to perform mass balances of micropollutants in sewage treatment plants. Application to pharmaceutical and personal care products (PPCPs). *Environmental Science & Technology*, *41*(3), 884–890. https://doi.org/10.1021/es061581g

[19] Rivas, A., Irizar, I., & Ayesa, E. (2008). Model-based optimisation of wastewater treatment plants design. *Environmental Modelling & Software*, *23*(4), 435–450. https://doi.org/10.1016/j.envsoft.2007.06.009

[20] Malaysia Sewerage Industry Guidelines (MSIG) - Sewage Treatment Plants, Volume 4 (2010).

[21] APHA, AWWA, & WEF. (2017). Standard methods for the examination of water and wastewater. In *Journal of Chemical Information and Modeling* (23rd ed.). https://doi.org/10.2105/SMWW.2882.219

[22] Kumar, P. R., Pinto, L. B., & Somashekar, R. (2010). Assessment of the efficiency of sewage treatment plants: a comparative study between Nagasandra and Mailasandra sewage treatment plants. *Kathmandu University Journal of Science, Engineering and Technology*, *6*(2), 115–125. https://doi.org/10.3126/kuset.v6i2.4020

[23] Bolong, N., Ismail, A. F., Salim, M. R., & Matsuura, T. (2009). A review of the effects of emerging contaminants in wastewater and options for their removal. Desalination, 239(1–3), 229-246. https://doi.org/10.1016/j.desal.2008.03.020

[24] Crites, R. W. (1994). Design criteria and practice for constructed wetlands. *Water Science and Technology*, *29*(4), 1–6. https://doi.org/10.2166/wst.1994.0144

[25] Dutta, A., & Sarkar, S. (2015). Sequencing batch reactor for wastewater treatment: Recent advances. *Current Pollution Reports*, *1*(3), 177–190. https://doi.org/10.1007/s40726-015-

0016-y
[26] Vigneswaran, S., Sundaravadivel, M., & Chaudhary, D. S. (2007). Sequencing batch reactors: Principles, design, operation and case studies. *Water and Wastewater Treatment Technologies*.
[27] DDM. (2018). *Design parameters of UMS-Central STP*.
[28] Choksi, K. N., Sheth, M. A., And, & Mehta, D. (2015). To assess the performance of Sewage Treatment Plant : A Case study of Surat city. *International Research Journal of Engineering and Technology*, *2*(8), 1071–1075.
[29] Singh Asiwal, R., Kumar Sar, S., Singh, S., & Sahu, M. (2016). Wastewater treatment by effluent treatment plants. *International Journal of Civil Engineering*, *3*(12), 19–24. https://doi.org/10.14445/23488352/IJCE-V3I12P105
[30] Abd Mutalib, N. A., Karim, O. A., & Mustafa, A. D. (2015). The water quality study and sources of pollution in Alur Ilmu, UKM. *Malaysian Journal of Analytical Sciences*, *19*(5), 1137–1146.
[31] Fan, L., & Xie, Y. (2011). Optimization control of SBR wastewater treatment process based on pattern recognition. *Procedia Environmental Sciences*, *10*(PART A), 20–25. https://doi.org/10.1016/j.proenv.2011.09.005
[32] Department of Environment Malaysia (DOE). "Environmental Quality Act report." Ministry of Science Technology and the Environment Putrajaya Malaysia (2012).
[33] Jamwal, P., Mittal, A. K. & Mouchel, J-M. (2009). Efficiency Evaluation of Sewage Treatment Plants with Different Technologies in Delhi (India). Environmental Monitoring and Assessment 153 (1-4). 293–305.
[34] Arborea, S., Giannoccaro, G., de Gennaro, B., Iacobellis, V., & Piccinni, A. (2017). Cost-Benefit Analysis of wastewater reuse in Puglia, Southern Italy. *Water*, *9*(3), 175. https://doi.org/10.3390/w9030175
[35] Ali, A. N. A, Khan, M. Y. M. A., Bolong, N., Maarof, K. A. & Tanalol, S. H. (2020). Conceptual and design framework for smart stormwater filtration. In 2nd International Conference on Artificial Intelligence in Engineering and Technology (IICAIET), 2020, pp. 1-4, doi: 10.1109/IICAIET49801.2020.9257828.
[36] Hidalgo, D., & Irusta, R. (2005). The cost of wastewater reclamation and reuse in agricultural production in the Mediterranean countries. *IWA International Conference on Water, Economics, Statistics and Finance, October 2014*, 1–8.
[37] Dawoud, M. A., Sallam, O. M., & Abdelfattah, M. A. (2012). Treated wastewater management and reuse in Arid Regions. *The 10th Gulf Water Conference 22-24, January*, 732–752.

Acknowledgments
The authors would like to acknowledge the financial support from the Ministry of Higher Education (FRGS/1/2020/TK0/UMS/02/3) and UMSGreat (GUG0441-1/2020) research grants. Thanks to the UMS Department of Development and Maintenance, especially Ir. M. Ari Nazrul A. Rahman and Ir. Dr. Asrul Effendy Ismail, for their assistance in this research. We also wholeheartedly thank the reviewer's input and helpful suggestions for the paper's improvement.

Quantification and Characterisation of Microplastics in Fish and Surface Water at Melayu River, Johor

A Primus[1] and S Azman[1]

[1]Department of Water and Environmental Engineering, School of Civil Engineering, Faculty of Engineering, Universiti Teknologi Malaysia, 81310 Johor Bahru, Johor, Malaysia.

Email: shamila@utm.my

Abstract. Microplastics are plastic particles (< 5 mm) found in the environment that can be ingested by animals and transferred up in the trophic level. The study was conducted through sample collection, digestion of gastrointestinal (GI) tracts of fish, density separation using NaCl, filtration, microscopy, and ATR-FTIR Spectroscopy. The amount of ingested microplastics by Melayu River fish samples was: Gray Eel-Catfish (3.92 ± 4.17 particles/individual) > Sagor Catfish (2.00 ± 1.41 particles/individual) > Spotted Sicklefish (2.00 ± 0.00 particles/individual). The trend of microplastics by month in water samples was Mar-20 (2.89 ± 1.36 particles/L) > Feb-20 (1.33 ± 1.00 particle/L) and Jan-20 (1.00 ± 0.87 particle/L). Microplastics were mostly in the class size 0 µm – 0.50 µm. In the fish samples, fibres were found to be dominant. In water samples, films were dominant. Ingestion of microplastics by colour was ranked as blue > black > red > yellow in fish samples whereas microplastics' colour in water samples was ranked as blue > red > black > translucent > green. Therefore, it is concluded that the abundance of blue microplastics in fish samples was due to the common blue plastics used by the locals. The identified microplastics were of polyethylene terephthalate (PET) and polyethene (PE) origins.

1. Introduction

Global plastic production increased from 359 million to 368 million from 2018 to 2019 wherein 2019, plastic production is divided into packaging (39.6 %), construction materials (20.4 %), automotive industries (9.6 %), electronic and electrical appliances (6.2 %), household, leisure and sports items (3.4 %) and 16.7 % in items in various industries (e.g., medical) (PlasticsEurope, 2020). Plastic polymers are often hard to biodegrade but if they do, they require the help of organisms that have degradation ability such as mealworms (Horton et al., 2017).

Global plastic production increased from 359 million to 368 million from 2018 to 2019 wherein 2019, plastic production is divided into packaging (39.6 %), construction materials (20.4 %), automotive industries (9.6 %), electronic and electrical appliances (6.2 %), household, leisure and sports items (3.4 %) and 16.7 % in items in various industries (e.g., medical) (PlasticsEurope, 2020). Plastic polymers are often hard to biodegrade but if they do, they require the help of organisms that have degradation ability such as mealworms (Horton et al., 2017).

Microplastics are plastic particles (<5 mm) that can travel long-distance by the circulation of the water or sea surface due to their buoyant properties (Density between 0.917-1.61 g/cm^{-3}) and prevalent nature (Campbell et al., 2017; Thevenon & Carroll, 2015; Wu et al., 2019). Plastics can be broken down into microplastics through exposure to UV radiation which causes the plastics to become brittle.

Microplastics are divided into primary microplastics which are released readily into the environment as small particles (e.g., microbeads in facial cleanser) and secondary microplastics that include fragmentation of larger plastics (e.g., packaging) (Weis, 2020; Wu et al., 2019). Microplastics can enter the aquatic environment through fishing activities, tourist ships, toiletries, and the washing of synthetic textiles (Karbalaei et al., 2018). Surface water is reported to have a higher amount of polystyrene (PS), polypropylene (PP), and polyethylene (PE) (Wu et al., 2019).

Freshwater microplastics have become increasingly studied (Ben et al., 2021; Campbell et al., 2017). These microplastics originate from land-based sources which are dependent on the level of urbanization, plastic disposal behaviour, and common plastic usage (Ben et al., 2021). Microplastics can become a vector to harmful substances such as heavy metals and persistent organic pollutants (POPs) present in the environment and can also leach out additives that were added during the plastic manufacturing process (Wu et al., 2019). Exposure to both microplastics and chemical contaminants is said to increase toxicity levels in several organisms (Ha & Yeo, 2018).

Microplastics tend to be ingested by aquatic animals, typically by fish as microplastics can look similar to their natural prey [10,11]. The amount of microplastics that enter freshwater organisms depends on their feeding behaviour. Microphagous animal is likely to accidentally ingest microplastics along with their natural food [12]. Benthic organisms will mostly encounter high-density microplastics on the bottom sediment [6]. For fish, ingestion of microplastics will obstruct and damage their digestive tracts and can cause them to starve to death [13]. Microplastics can travel up the trophic level via the prey-predator relationship [14] and will eventually end up in the human digestive system.

The findings in this study determined the amount of ingested microplastics in common fish found in the Melayu River in Malaysia as well as the microplastic abundance in its surface water. Additionally, the findings helped to highlight safety in food consumption and the consequences of water pollution. Humans consumed up to 5 g of plastic every week due to incidental plastic intake when consuming food and drink [15]. According to several studies, fibres were most dominant in freshwater fish as fibres' shapes are flexible and easier to be ingested [16,17]. The colour of microplastics ingested by fish differs by species. Some of the species rely on visual to consume food which can be affected by the level of illumination in the aquatic environment whereas other relies on specialized chemical sense regulators to detect potential food [17]. Although this study focused only on the gastrointestinal (GI) tracts of fish, other parts of the fish such as gills can also accumulate microplastics [18]. In Johor, Malaysia, there is still a lack of studies on microplastics, particularly in surface water. In Skudai and Tebrau rivers, the mean microplastics in sediment were 200 ± 80 particles/kg and 680 ± 140 particles/kg respectively. However, sediment acts as a sink for microplastics and therefore contained a higher amount of microplastics [19,20]. In addition, both rivers were located in a highly-populated area as opposed to the Melayu River.

2. Methodology

This section detailed the methodology for this study. Methodology used in this study included site selection, collection and dissection of fish, collection and filtration of water samples, verification of microplastics and analysis of data.

2.1. Site selection

Melayu River (1°27'20.8"N, 103°41'31.3"E) is located in Johor Bahru, Johor, Malaysia. The river starts from the residential area in Nusajaya (Horizon Hills) and flows southeast towards Kampung Sungai Melayu. The mouth of the river is located at Kampung Sungai Melayu and the river water discharges towards the Straits of Johor. A microplastic study was conducted as the Melayu River is susceptible to various pollutants due to its vicinity to residential establishments. Additionally, the river is the main source of income for local fishermen residing in Kampung Sungai Melayu. The village itself is surrounded by a vast mangrove forest which is a stopover for migratory birds. The river was reserved for eco-tourism activities. Among the activities include a tour along the river, crab catching, bird watching, fishing, and handicrafts. Figure 1 and Table 1 show the sampling locations.

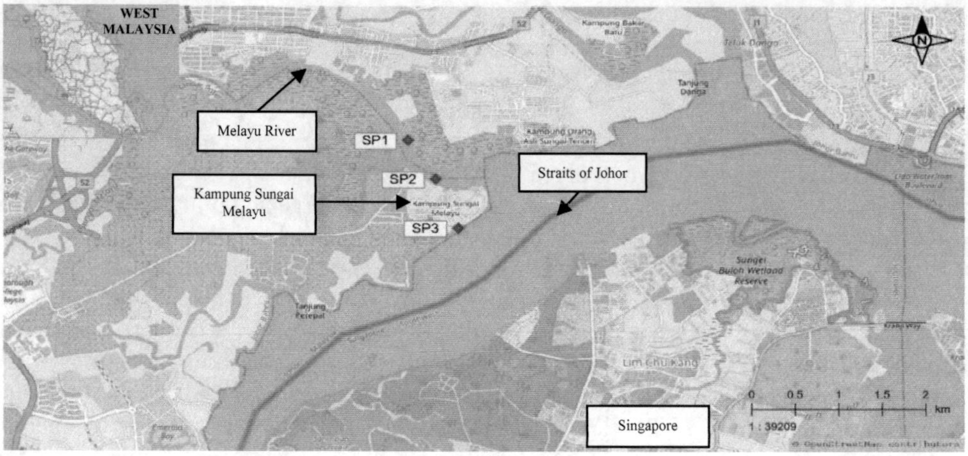

Figure 1. Sampling locations along Melayu River.

Table 1. Coordinates of the sampling locations.

Label	Location	Significance	Latitude	Longitude
SP1	Port Memancing Kampung Sungai Danga	Upstream	1° 27' 35.9" N	103° 41' 20.0" E
SP2	Restoran Sri Pantai Asam Pedas Kampung Sungai Melayu	Middle	1° 27' 19.3" N	103° 41' 31.0" E
SP3	Kampung Sungai Melayu	Downstream	1° 26' 58.8" N	103° 41' 39.3" E

2.2. Collection and dissection of fish

Spotted Sicklefish (Drepane punctata), Sagor Catfish (Hexanematichthys sagor), and Gray Eel-Catfish (Plotosus canius) are some fish species that inhabit the study location. They are both eaten and sold at the local wet market. Sagor Catfish and Gray Eel-Catfish are amphidromous [21] which means species that migrate from freshwater to sea and vice-versa whereas Spotted Sicklefish can be found in shallow coastal water [22]. Additionally, Spotted Sicklefish is described as amphidromous in the FishBase database.

The fish were wild-captured on 6th January, 20th February, and 9th March 2020. They were kept in a basin with some in-situ water and transported back to the Environmental Laboratory of Universiti Teknologi Malaysia (UTM). The fish body weight (g) and total length (from snout to tail) (cm) were recorded before carefully removing their GI tracts using a knife and kitchen scissors. The GI tract (oesophagus, stomach, and intestine) was isolated carefully from other tissues, weighted, stored in a petri dish, and refrigerated until further study.

The method of extraction of the GI tract of fish followed [23] with modifications. 10% potassium hydroxide (KOH) was used for the digestion of GI tracts. To obtain 10% KOH, 100 g KOH pellets were weighed and diluted in a beaker using distilled water up to the 1000 mL mark. GI tracts were put into separate pre-labelled 250 mL conical flasks and added with 10 % KOH up to 3 times the volume of GI tracts. The conical flasks were covered with aluminium foil to avoid airborne contamination. The GI tracts were left to digest for 72 hours at room temperature to prevent heat damage to microplastics. After digestion, NaCl solution (Density 1.2 g/cm-3) was used to float microplastics in the solution. The conical flasks were gently swirled and left to settle overnight. Then, the supernatant formed was filtered through a vacuum pump using filter papers with 22 μm pore size (Grade 541, Ø 47 mm, Whatman®). The moist filter paper was oven-dried at 40 °C until a constant weight was achieved.

2.3. *Collection and filtration of water samples*
3 of 1 L of surface water samples were collected at each sampling location on the same day as fish specimen collection. Until further steps, the water samples were stored in a dark cold room at 4°C [24] in the Environmental Laboratory of Universiti Teknologi Malaysia (UTM) to preserve the water samples and the fish samples were stored in a refrigerator. The water samples were filtered through a vacuum pump using filter papers with 22 μm pore size (Grade 541, Ø 47 mm, Whatman®). The moist filter paper was oven-dried at 40 °C until a constant weight was achieved [24].

2.4. *Visual inspection of microplastics*
Microplastics were identified using a stereomicroscope (Model: HSZ-600) after the filtration process. The magnifications were set from 10x - 40x. Particles of organic compounds and tiny grains of sand were carefully removed. Particles that were untearable by tweezers were identified as potential microplastics and were measured using *i-solution* Premium software. 'Hot needle' test was also used to confirm suspected microplastics [4]. The microplastics obtained were classified based on their shape and colour. The colour definitions were black (including grey), red (including pink), yellow (including orange), blue (including green), white and translucent. Microplastics are divided into 5 shapes; film, fibre, fragment, bead, and foam. As microplastics are plastic particles of size 5 mm and less, i.e., plastic particles larger than 5 mm have been omitted from this study.

2.5. *Verification of microplastics*
ATR-FTIR analysis was conducted to verify and characterize microplastics in the samples as it reduces false-positive data by using chemical confirmation as opposed to using only a microscope [25]. The microplastics particles were transferred into a clean petri dish with new filter paper and analysed between wavelengths of 650 - 4000 cm^{-1} [26].

2.6. *Analysis of data*
The abundance of microplastics in fish samples (Mean ± SD and percentage) in terms of quantity, class size, shape, and colour were plotted using OriginPro 2021. Similarly, the result for microplastic abundance in water samples was plotted using the software. The functional groups of microplastics were determined by comparing the ATR-FTIR results with past studies [27].

3. Results and discussion
This section detailed the results obtained from this study and also the discussion on the acquired data.

3.1. *Physical parameters of fish species*
Microplastics are either ingested by fish and other animals and might sink to the bottom and be wedged between the sediments. The total amount of microplastics ingested by Spotted Sicklefish (*Drepane punctata*) were 2 particles whereas Sagor Catfish (*Hexanematichthys sagor*) and Gray Eel-Catfish (*Plotosus canius*) ingested a total amount of 4 and 47 particles respectively. The abundance of microplastics (Mean ± SD) in each fish species were 2.00 ± 0.00 particles/individual, 2.00 ± 1.41 particles/individual, and 3.92 ± 4.17 particles/individual respectively. This was because only a small number of Spotted Sicklefish and Sagor Catfish were able to be captured during the three (3) months of sampling whereas Gray Eel-Catfish can be found abundant in the area. The amount of ingested microplastics ranged from 0 - 9 particles/individual. Sagor Catfish ingested more microplastics than Spotted Sicklefish due to their larger GI tracts.

The total microplastics in sampling 1 (Jan-20), 2 (Feb-20), and 3 (Mar-20) were 9, 12 and 26 particles respectively and the range of microplastics in water samples was 0 - 6 particles/L. The abundance of microplastics (Mean ± SD) in the water samples were 1.00 ± 0.87 particle/L, 1.33 ± 1.00 particle/L and

2.89 ± 1.36 particles/L respectively. Water samples collected in March 2020 has the highest number of microplastics. According to the fishermen in the area, the mating season of mussels occurred during March. Thus, the burying behaviour of the mussels may have contributed to the turbid water during March 2020 and helped to dislodge microplastics from sediment. Microplastics can resuspend on the surface water due to disturbances of sediment by benthic organisms' activities [4]. The number of microplastics in fish specimens was higher than in the water samples. The presence of biofilm on the microplastics' surfaces may have sent chemical signals to trigger the feeding response in fish [14].

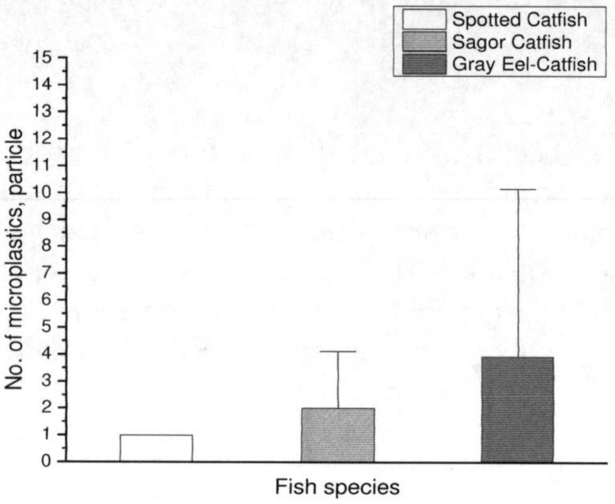

Figure 2. Number of microplastics ingested by fish specimens.

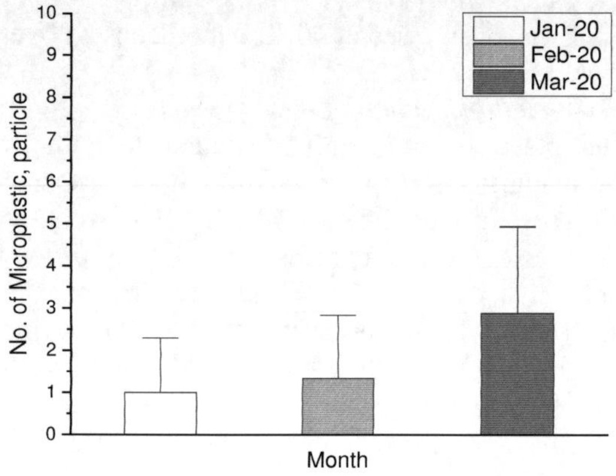

Figure 3. Number of microplastics in water samples.

3.2. Size of microplastics

The microplastics identified ranged from 0.22 μm - 3220 μm. The microplastics that were of size 0 μm - 0.50 μm were recorded as the highest (Spotted Sicklefish, *n*=0; Sagor Catfish, *n*=0; Gray Eel-Catfish, *n*=23) at 0 %, 0 % and 100 % respectively (refer Figure 4). Smaller particles are easier to be accidentally ingested than bigger particles [14]. Meanwhile, microplastics that were of size 1.51 μm -2.00 μm were the lowest (Spotted Sicklefish, *n*=0; Sagor Catfish, *n*=2; Gray Eel-Catfish, *n*=1) at 0 %, 67 % and 33 % respectively. The abundance of microplastics that were of size 0.51 μm – 1.00 μm in Spotted Sicklefish, Sagor Catfish, Gray Eel-Catfish were 0 % (*n*=0), 0 % (*n*=0) and 100 % (*n*=10) and of size 1.01 μm – 1.50 μm were 0 % (*n*=0), 0 % (*n*=0) and 100 % (*n*=7). Microplastics that were of size 2.00 μm and above were determined to be 20 % (*n*=2), 20 % (*n*=2) and 60 % (*n*=6) respectively.

In water samples, the microplastics identified range from 0.35 μm - 3410 μm.

Figure 5 shows that the majority of the microplastics found in water samples were in the size class of 0 μm - 0.50 μm (Jan-20, *n*=4; Feb-20, *n*=5; Mar-20, *n*=8) at 24 %, 29 % and 47 % respectively. Similar to the result for fish samples, the least abundant microplastics were found in the class size 1.51 μm -2.00 μm (Jan-20, *n*=0; Feb-20, *n*=1; Mar-20, *n*=1) at 0 %, 50 % and 50 % respectively. Meanwhile, the abundance of microplastics that were of size 0.51 μm – 1.00 μm (Jan-20, *n*=1; Feb-20, *n*=5; Mar-20, *n*=10) at 6 %, 31 % and 63 % respectively and of size 1.01 μm – 1.50 μm were 33 % (*n*=3), 11 % (*n*=1) and 56 % (*n*=5). Microplastics that were 2.00 μm and above were 33 % (*n*=1), 0 % (*n*=0) and 67 % (*n*=2) respectively. The microplastics in fish samples were similar in size to those in surface waters. This may be due to the sinking and resuspension of microplastics between sediment and surface water.

There are no existing standardized size criteria as size is a very common parameter in microplastic studies. It can be said that every research journal has presented a different class size, making it difficult to cite. Although plastic particles undergo mild fragmentation in freshwater settings than marine settings due to less wave action and turbulence, they will still undergo thermal and photo-oxidation by UV exposure to become smaller particles [12].

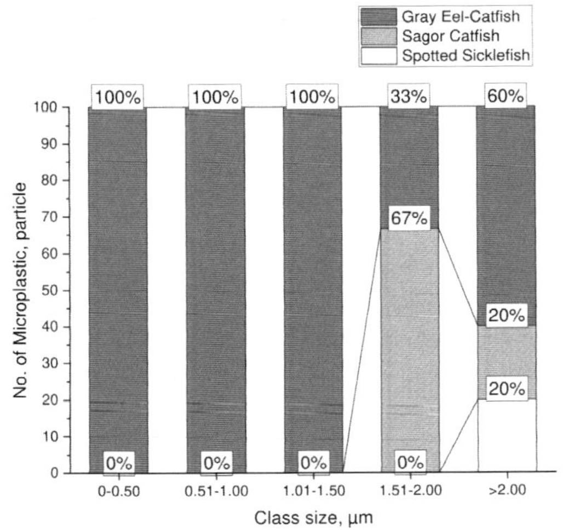

Figure 4. Size category of microplastics in fish species.

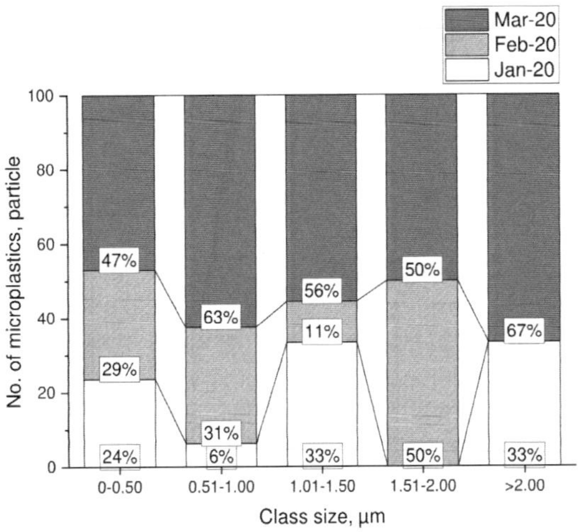

Figure 5. Size category of microplastics in water samples.

3.3. Shape of microplastics

Based on Figure 6, the percentage of occurrence of microplastic was 100 % fibres for Spotted Sicklefish, 100% fibres for Sagor Catfish. In Gray Eel-Catfish, the numbers of films found were higher than that of fibres at 47 % fibres and 53 % films.

Films originated from the degradation of plastic bags [28]. The presence of films indicates poor waste management. From Figure 7, the percentage of occurrence was 56 % fibres and 44 % films for Jan-20, 33 % fibres and 67 % films for Feb-20 and 12 % fibres and 88 % films for Mar-20. Figure 8 shows the shapes of microplastics present in the samples which consists of different coloured fibres and films.

Low-density microplastics can be present in the sediment zone due to biofouling. Biofouling causes the accumulation of foulants on the microplastics' surface which increase their density and cause them to sink and settle in the sediment layer [12]. Fibres are the most common microplastic type observed in sediments [7,27]. These fibres can be resuspended due to several reasons such as weather, bioturbation, tidal forcing, and trawling activity [29].

Fibres may originate from the use of fishing equipment as fishing is the main source of economy of the area. As the area of study is located near residential areas, fibres may also come from domestic effluent [28]. More than 100 fibres can originate from 1 L of wastewaters from washing synthetic clothes and an average of 1900 microfibres can be released in a single machine wash [13].

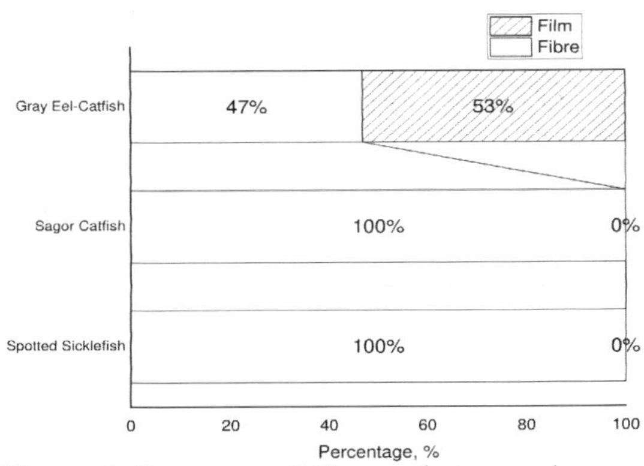

Figure 6. Percentage difference between shapes of microplastics ingested by fish species.

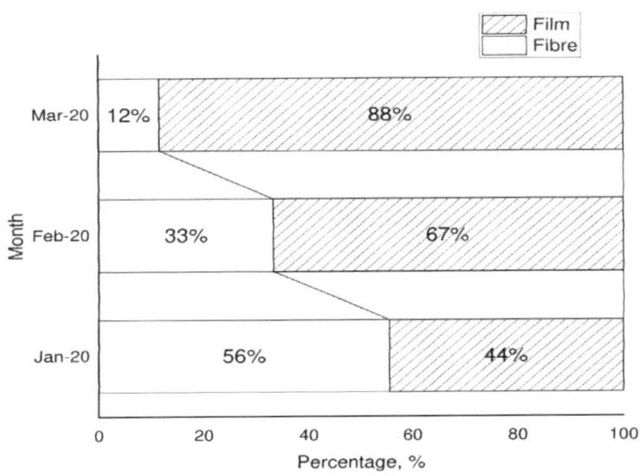

Figure 7. Percentage difference between shapes of microplastics in water samples.

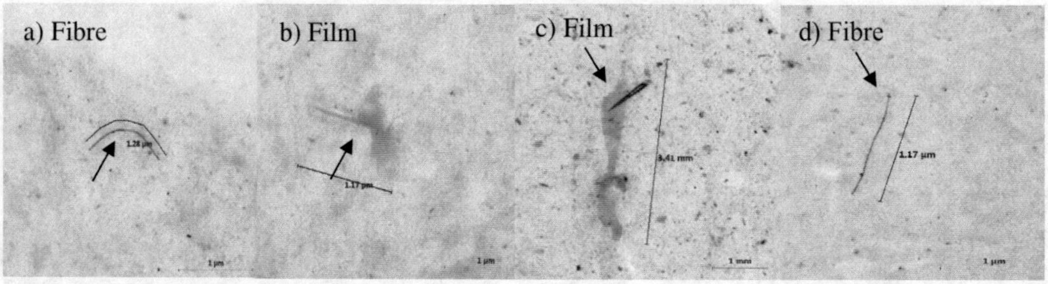

Figure 8. Microplastic in samples at 40x magnification: (a), (b) in fish; (c), (d) in surface water.

3.4. *Colour of microplastics*

Referring to Figure 9, the fish samples ingested mostly blue microplastics at 45 % (*n*=24) and the least amount of microplastics in yellow colour at 2 % (*n*=1). Amberstripe scads (*Decapterus muroadsi*) showed a preference for blue microplastics as the microplastics resembled their common prey which is the blue-coloured copepod species. This suggests that fish mistakenly ingested microplastics that resemble their natural prey [30].

In water samples (refer Figure 10), there is a very narrow scope for comparison with other worldwide researches since the microplastic colour depends on the majority of the colour of plastic used in the study area. Based on Figure 10, blue microplastics were determined to be the most abundant at 49 % (*n*=23). It is safe to assume that most residents in the study area used blue-coloured plastics in their daily routines followed by black-coloured plastics, red-coloured plastics, green-coloured plastics and transparent plastics. Thus, it can also be concluded that the abundance of ingested blue microplastics in fish samples came from surface water.

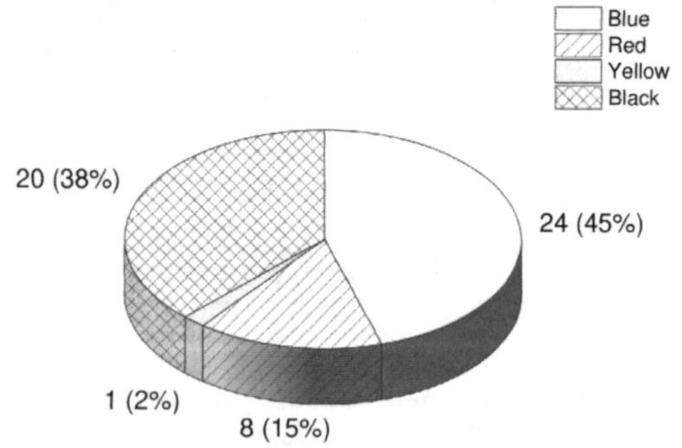

Figure 9. Colours of microplastics ingested by fish species.

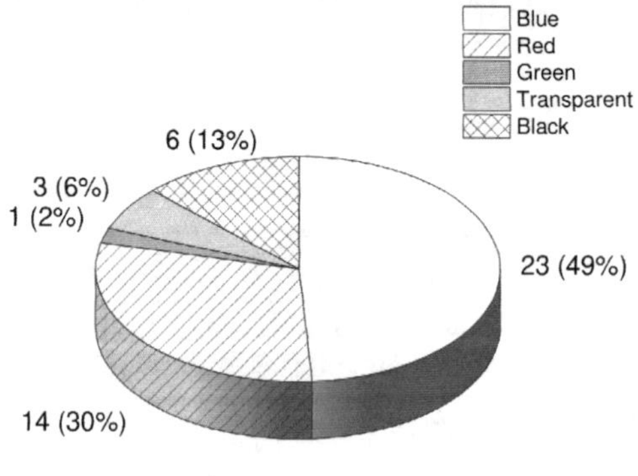

Figure 10. Colours of microplastics in water samples.

3.5. *3.6 ATR-FTIR analysis*

Spectra produced from the microplastics were compared with a study by Jung et. al [31]. The adsorption bands used for identifications provide an almost perfect match to polyethene (PE) and polyethylene terephthalate (PET). Fibres were determined to be PET whereas films were determined to be PE. PET and PE are common polymers used in the textile and packaging industry respectively [32,33]. Thus, domestic effluent and improper plastic waste management can be pinpointed to the sources of the microplastics.

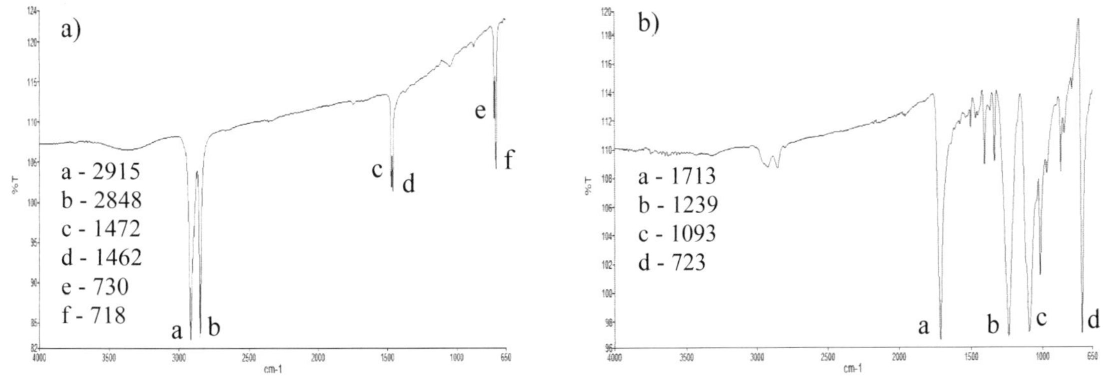

Figure 11. Spectra produced from microplastics: a) PE; b) PET

4. Conclusion

Microplastics were found in both Spotted Sicklefish and Sagor Catfish, but only 6 out of 12 Gray Eel-Catfish species. All water samples contained microplastic which indicates that microplastic contamination is a problem in the study area. There were significant variations in the number of microplastics between Gray Eel-Catfish and the other two fish species due to differences in the number of specimens and GI tract size.

The water samples taken during March 2020 contained more microplastics due to possible bioturbation during mussels mating season. Overall, there were more fibres in fish species and more films in water samples. Microplastics can sink and resuspend in the water-sediment zone.

The abundance of blue microplastics in the water samples may be due to the typical blue plastic bags used in the study field and the preferences of fish towards different colour microplastics suggest accidental ingestion towards microplastics similar to their natural prey. It is safe to assume that the surface water contributes to the abundance of ingested microplastics by the fish samples. Unfortunately, the exact source of these microplastics is unknown. However, they are most likely to be present due to improper plastic waste disposal, domestic effluent that flows into the river body, and the fishing activities in the study area. Further study should be done to assess microplastic content in different organs of fish to investigate the translocation of microplastics in fish which may impose harm to humans upon consumption. Overall, the microplastic pollution level can be assumed to be at a low level. However, it is always important to practice proper plastic waste disposal and be knowledgeable in what food we consume as microplastics can bioaccumulate in aquatic animals.

References

[1] PlasticsEurope 2020 Plastics – the Facts 2020: an analysis of European plastics production, demand and waste data *PlasticEurope*

[2] Horton A A, Walton A, Spurgeon D J, Lahive E and Svendsen C 2017 Microplastics in freshwater and terrestrial environments: Evaluating the current understanding to identify the knowledge gaps and future research priorities *Science of the Total Environment* **586** 127–41

[3] Thevenon F and Carroll C 2015 *Plastic debris in the ocean: the characterization of marine plastics and their environmental impacts, situation analysis report*

[4] Campbell S H, Williamson P R and Hall B D 2017 Microplastics in the gastrointestinal tracts of fish and the water from an urban prairie creek *Facets* **2** 395–409

[5] Wu P, Huang J, Zheng Y, Yang Y, Zhang Y, He F, Chen H, Quan G, Yan J, Li T and Gao B 2019 Environmental occurrences, fate, and impacts of microplastics *Ecotoxicology and Environmental Safety* **184** 109612

[6] Weis J S 2020 Aquatic Microplastic Research—A Critique and Suggestions for the Future *Water* **12** 1475
[7] Karbalaei S, Hanachi P, Walker T R and Cole M 2018 Occurrence, sources, human health impacts and mitigation of microplastic pollution *Environmental Science and Pollution Research* **25** 36046–63
[8] Ben P, Andreou; D, Green; L D and Britton J R 2021 Microplastics in freshwater fishes: Occurrence, impacts and future perspectives *Fish and Fisheries* **22** 467–88
[9] Ha J and Yeo M-K K 2018 The environmental effects of microplastics on aquatic ecosystems *Molecular and Cellular Toxicology* **14** 353–9
[10] Qi H, Fu D, Wang Z, Gao M and Peng L 2020 Microplastics occurrence and spatial distribution in seawater and sediment of Haikou Bay in the northern South China Sea *Estuarine, Coastal and Shelf Science* **239** 106757
[11] Wright S L, Thompson R C and Galloway T S 2013 The physical impacts of microplastics on marine organisms: a review. *Environmental pollution (Barking, Essex : 1987)* **178** 483–92
[12] Wong J K H, Lee K K, Tang K H D and Yap P S 2020 Microplastics in the freshwater and terrestrial environments: Prevalence, fates, impacts and sustainable solutions *Science of the Total Environment* **719** 137512
[13] Espinosa C, Esteban M Á and Cuesta A 2016 Microplastics in Aquatic Environments and Their Toxicological Implications for Fish *Toxicology - New Aspects to This Scientific Conundrum*
[14] Ašmonaitė G and Almroth B C 2019 *Effects of Microplastics on Organisms and Impacts on the Environment: Balancing the Known and Unknown*
[15] Ribeiro F, Okoffo E D, O'Brien J W, Fraissinet-Tachet S, O'Brien S, Gallen M, Samanipour S, Kaserzon S, Mueller J F, Galloway T and Thomas K V. 2020 Quantitative Analysis of Selected Plastics in High-Commercial-Value Australian Seafood by Pyrolysis Gas Chromatography Mass Spectrometry *Environmental Science and Technology* **54** 9408–17
[16] U O and M V 2017 Occurrence and Impacts of Microplastics in Freshwater Fish *Journal of Aquaculture & Marine Biology* **5** 1–4
[17] Sarijan S, Azman S, Said M I M and Jamal M H 2021 Microplastics in freshwater ecosystems: a recent review of occurrence, analysis, potential impacts, and research needs *Environmental Science and Pollution Research* **28** 1341–56
[18] Kuśmierek N and Popiołek M 2020 Microplastics in freshwater fish from Central European lowland river (Widawa R., SW Poland) *Environmental Science and Pollution Research* **27** 11438–42
[19] Maes T, van der Meulen M D, Devriese L I, Leslie H A, Huvet A, Frère L, Robbens J and Vethaak A D 2017 Microplastics baseline surveys at the water surface and in sediments of the North-East Atlantic *Frontiers in Marine Science* **4**
[20] Bellasi A, Binda G, Pozzi A, Galafassi S, Volta P and Bettinetti R 2020 Microplastic Contamination in Freshwater Environments: A Review, Focusing on Interactions with Sediments and Benthic Organisms *Environments* **7** 30
[21] Zobkov M B and Esiukova E E 2018 Microplastics in a Marine Environment: Review of Methods for Sampling, Processing, and Analyzing Microplastics in Water, Bottom Sediments, and Coastal Deposits *Oceanology* **58** 137–43
[22] Shim W J, Hong S H and Eo S E 2017 Identification methods in microplastic analysis: a review *Analytical Methods* **9** 1384–91
[23] de Falco F, di Pace E, Cocca M and Avella M 2019 The contribution of washing processes of synthetic clothes to microplastic pollution *Scientific Reports* **9** 1–11
[24] Li Y, Lu Z, Zheng H, Wang J and Chen C 2020 Microplastics in surface water and sediments of Chongming Island in the Yangtze Estuary, China *Environmental Sciences Europe* **32**
[25] Yuan W, Liu X, Wang W, Di M and Wang J 2019 Microplastic abundance, distribution and composition in water, sediments, and wild fish from Poyang Lake, China *Ecotoxicology and Environmental Safety* **170** 180–7

[26] Martin J, Lusher A, Thompson R C and Morley A 2017 The Deposition and Accumulation of Microplastics in Marine Sediments and Bottom Water from the Irish Continental Shelf *Scientific Reports* **7** 1–9
[27] Ory N C, Sobral P, Ferreira J L and Thiel M 2017 Amberstripe scad Decapterus muroadsi (Carangidae) fish ingest blue microplastics resembling their copepod prey along the coast of Rapa Nui (Easter Island) in the South Pacific subtropical gyre *Science of the Total Environment* **586** 430–7

Performance of Ozone/$SnCl_4$ Oxidation in Stabilized Landfill Leachate Treatment

S N F Zakaria[1*], N Bolong[1], I Saad[1] and H A Aziz[2,3]
[1]Faculty of Engineering, Universiti Malaysia Sabah, Jalan UMS, 88400 Kota Kinabalu, Sabah, Malaysia.
[2]School of Civil Engineering, Engineering Campus, Universiti Sains Malaysia, 14300 Nibong Tebal, Penang, Malaysia.
[3]Solid Waste Management Cluster, Science and Technology Research Centre, Engineering Campus, Universiti Sains Malaysia, 14300 Nibong Tebal, Penang, Malaysia.

*Corresponding author: snfarhanazakaria@ums.edu.my

Abstract. Landfill leachate is well known as contains a high concentration of pollutants and threatens our environment as well as human life. Therefore, this study investigates the potential of ozonation (O_3) and tin tetrachloride ($SnCl_4$) in remedying stabilised landfill leachate (SAL). The performance of this treatment is evaluated by measuring the percentage removal of chemical oxygen demand (COD) and colour at different dosages, pH, and reaction times. The optimum percentage removals recorded for colour and COD were 98% and 83.2%, respectively, at an optimum dosage ratio (COD: $SnCl_4$) 1:1.25, reaction time of 60 min and pH 8.5. Moreover, the biodegradability ratio of the sample also shows an improvement from 0.03 to 0.28, using this treatment. Thus, these data show that the oxidation process has a great potential to remediate recalcitrant pollutants like leachate

Keywords: Stabilised leachate, ozonation, advance oxidation process, biodegradability, tin tetrachloride

1. Introduction

The amount of solid waste and leachate production in Malaysia is increasing every day. On top of that, landfill by-products, especially in anaerobic stabilised leachate, contains low biodegradability and high toxicity levels which can be harmful to the environment and human beings. Low biodegradability of leachate restrains the degradation of organic matter through biological processes. Organic matter such as ammonia may contribute to adverse impacts on health and the environment. Therefore, an effective treatment to treat leachate is necessary. Various studies have been conducted on anaerobic stabilised landfill leachate treatment and yet, it is still difficult to comply with the standard discharge limits of effluent. In addition, biological treatment method also inefficient for remediating SAL as this aging leachate contains a low quantity of oxygen which makes aerobic or anaerobic digestion is hard to occur. Thus, organic matter present in the waste will become more concentrated and complex as the days go by. Another prevalent treatment to remediate SAL is through chemical or physico-chemical treatment.

As seen in Figure 1, AOP have ability to degrade the complex structures into simpler structure using oxidative agent namely free radicals (i.e. hydroxyl) [1]. Ozonation is a part of AOP and have ability to remove pollutant effectively from the wastewater. However, ozonation alone is unable to remove all recalcitrant pollutants from the leachate. Several researchers recorded that O_3 alone displayed a poor performance compared to ozone/fenton and ozone/persulfate [2]. Lack amount of OH• in the solution make ozonation process unable to degrade the pollutant effectively compared to the O_3/catalyst. Enhancement of OH• generation using potential organic or inorganic compound in ozonation process have been studied widely by many scholars and summarised in Table 1. There are several disadvantages of the listed treatment such as precipitation and iron residuals. Therefore, a suitable element or catalyst is required in order to enhance the effectiveness of AOP in leachate treatment.

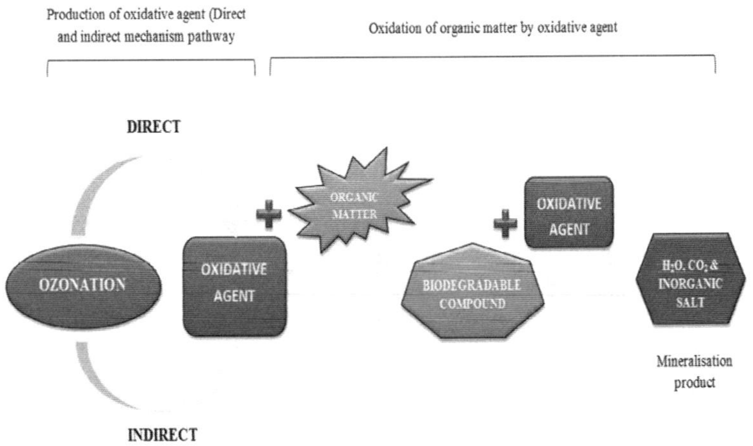

Figure 1. General mechanisms of ozonation in remediation of wastewater

Table 1. List of O_3/catalyst AOP in water and wastewater treatment

	AOP	Percentage removal	References
1.	Ozone/persulfate ($S_2O_8^{2-}$) oxidation process	COD: 72%, Colour: 93%, NH_3-N: 55%	[3]
2.	Ozone/catalyst chips	BY87: 99%, COD: 60%, TOC: 30%	[4]
4.	Ozone/ Zin Suplhate ($O_3/ZnSO_4^{2-}$)	COD: 90%, Colour: 99%	[5]
5.	Ozone/Sludge catalyst	Turbidity:13.02%, EC:10.57%, COD: 49.44%	[6]

In this study, tin tetrachloride or stannic chloride is selected as a potential element in enhances the effectiveness of ozonation process. $SnCl_4$ is categorised as an inorganic compound which has ability to accept electron pair donation and make this compound as a good catalyst or co-catalyst, especially in synthesis the organic matter. Moreover, the interaction of $SnCl_4$ and an aqueous solution resulted as hydrolysis, may increases the amount of OH ions in an aqueous solution. Under this reaction, the decomposition of O_3 will be accelerated and increase the production of OH• as well as AOP performance [7]. Therefore, $SnCl_4$ has high potential as a good catalyst in ozonation process. Additional, $SnCl_4$ is also known as a transition metal, cheap Lewis acid and low toxicity [8]. The application of $SnCl_4$ in leachate treatment has been reported by many scholars. For example, [9] have used $SnCl_4$ as a coagulant to treat stabilised landfill leachate and recorded high percentage removals of COD, colour and suspended solid as 71.53%, 98.40% and 99.54%, respectively, however, the application of $SnCl_4$ in AOP is not well documented yet.

The aims of this study are to investigate the potential of $SnCl_4$ in ozonation process ($O_3/SnCl_4$) as a leachate treatment and to compare with O_3 alone. Hence, the influences of pH, dosage and reaction time were tested, plotted and evaluated in this study. In this test, the removal of colour and COD is used as measurement of the treatment performance. Afterward, the effect of both methods on biodegradability was tested and analysed in the current work. Additionally, the procedure involved in this study is listed in section 3 (materials and methods) and the discussion on the data obtained is done in section 4 (result and discussion) while the conclusions as well as recommendations for future studies is made in section 5 (conclusion).

2. Materials and methods

2.1. Leachate sampling and characteristics

Alor Pongsu landfill site (APLS) which is located in Alor Pongsu, Perak, Malaysia was selected as the case study. The samples were collected and transferred to the laboratory for analysis. The sampling procedures, preservation and analysis of sample were conducted in accordance with the Standard Methods for the Examination of Water and Wastewater [10].

2.2 Oxidation system

The ozonation process flow and equipment applied in this study are shown in Figure 2. A 500 mL stainless steel laboratory-scale reactor is custom made with an inner diameter of 8 cm and a height of 20 cm. The system is connected to a water bath to maintain the system temperature at 15 °C, and below [11]. The details instruments used is depicts in Table 2 [12].

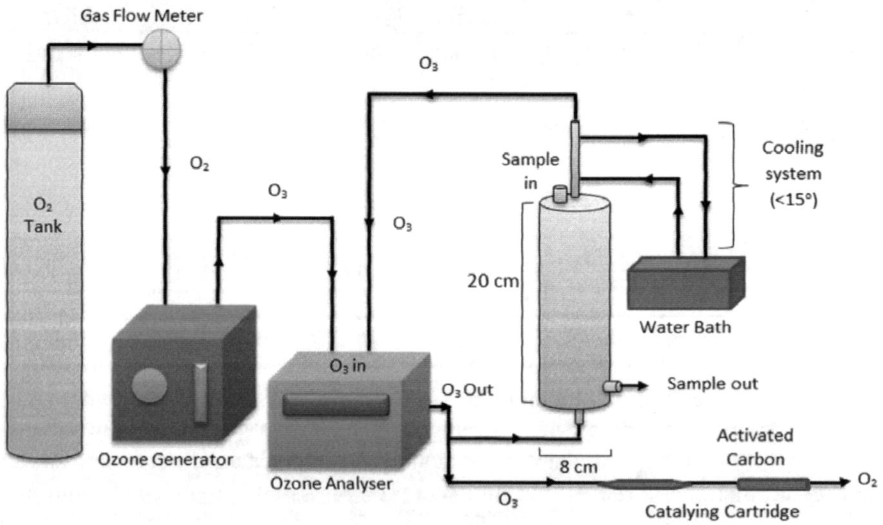

Figure 2. Schematic diagram of AOP equipment and experiment procedures.

2.3 Batch study

In the current study, the dosage, pH, and reaction time were varied to evaluate the influence of those factorials in O_3 and $O_3/SnCl_4$ processes as pH 5, 6, 7, 8, 9 and 10 while the reaction time ranged from 10, 30, 60, 90, 120 and 150 min, accordingly. In this study, $SnCl_4$ was and prepared as a 200 g/L stock solution by dissolving $SnCl_4$ chunk with 1 L of distilled water. After that, $SnCl_4$ stock solution was stored in a dark glass bottle to avoid unnecessary reactions, and the procedures are consistent with previous scholarly works [13]. Subsequently, the required amount of $SnCl_4$ was mixed with SAL, and any adjustment of pH (if required) was made using hydrochloric acid (HCl) and sodium hydroxide (NaOH). The gas flow rate and sample volume were fixed at 2 L/min and 500 mL, respectively [14]. Afterward, the effluent of the ozonation process was analysed in triplicate for colour and COD removal.

Table 2. Summary of instruments used in this study

No	Name of Instrument	Model
1	Ozone generator	OG-001 (UTM, Malaysia)
2	Ultraviolet gas ozone analyser	BMT 964 (Messtechnik, Germany)
3	Ozone Reactor	Custom fabrication (Figure 2)
4	Water bath	NESLAR
5	pH meter	Mettler Toledo
6	DR 2800 spectrophotometer (Colour and COD analyser)	HACH
7	COD reactor	HACH

3. Result and discussion

3.1 Influence of pH

pH plays an important role in every chemical reaction. In order to understand the influence of pH on ozonation and removal, a few sets of experiments were designed. The experiments started with a control test (i.e., without ozonation or jar test), followed by the batch study test. Results from both tests were analysed and compared. A control study is important to understand if pH agents (HCl and NaOH) have an effect on the concentration of pollutants. Therefore, a control study was run, and the result is presented in Figure 3.

As presented in Figure 3, the colour concentration and COD decrease considerably at pH levels 3 and 4 (see Figure 3). Almost 82 % and 43 % of removal by HCl alone are recorded for colour and COD respectively at pH 3, while 45% and 23% of colour and COD removal are obtained at pH 4. Observation using the naked eye shows that the colour has changed from dark brown (raw sample) to light brown after pH adjustment. The unstable solution made by the adjustment to pH 3 and 4 exhibits a significantly different colour concentration from raw leachate. This result might be due to the fact that organic compounds are oxidised at extreme pH levels, and the removal at this pH is not because of the AOP reaction. Therefore, the results obtained at pH 3 and 4 are negligible in this study, and the pH only ranges from pH 5 to 10.

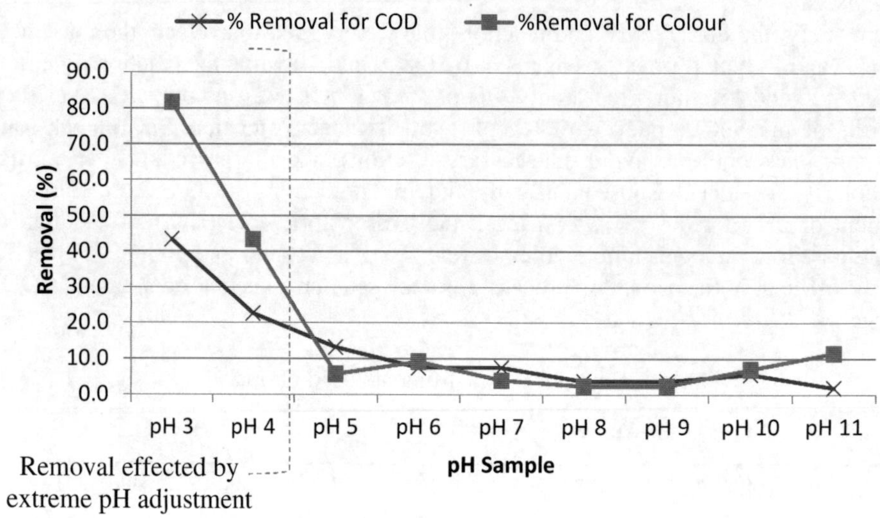

Figure 3. Control test without ozonation process

Figure 4 demonstrates that organic matter is susceptible to oxidation under alkaline conditions (i.e., pH >8) as the colour removal increases with pH value and starts to stabilise at pH 8 with 85% removal for O_3 and 98.2% for $O_3/SnCl_4$. As seen in Figure 4, COD also shares a similar pattern like colour, recording the highest percentage removal of COD of 50.1% for O_3 and 83.5% for $O_3/SnCl_4$ at alkaline conditions (pH 8 to pH 9), respectively. The tendency of O_3 and $O_3/SnCl_4$ to perform well under alkaline conditions shows that the indirect reaction mechanism is dominant in this treatment. The alkaline solution contains a high amount of OH ions which enhances the decomposition of O_3 (Equation 1) as well as accelerate the production of OH• (Equation 2 and 3) and increases the degradation of pollutant inside the sample [15][16]. Therefore, high removal of pollutants has been recorded at this pH condition.

$$O_3 + OH^- \rightarrow \bullet\ O_2^- + HO_2 \bullet \quad (1)$$

$$O_3 + \bullet\ O_2^- \rightarrow \bullet\ O_3^- + O_2 \quad (2)$$

$$\bullet\ O_3^- + H^+ \rightarrow \bullet\ HO_3 \rightarrow \bullet\ OH + O_2 \quad (3)$$

A comparison between ozonation alone and $O_3/SnCl_4$ show that $SnCl_4$ has improved the oxidation process and increased the percentage removal of colour and COD. The result may happen because of the hydrolysis of $SnCl_4$ with water (Equation 4) which increases the amount of OH ion in the solution and accelerates the decomposition of O_3 as well as OH• generation [17]. Additional, OH^-, which is originally present in aqueous conditions, also contributes to the decomposition of O_3 and generates other free radicals such as superoxide ion radicals and hydroperoxide ion radicals.

$$SnCl_4 + H_2O \rightarrow 2H_2\,[SnCl_6] + Sn(OH)_4 \quad (4)$$

Meanwhile, in low pH conditions, the removal of colour and COD are also detected here but not as high as in alkaline condition. This data indicates that the role of direct pathway reaction also occurs and ozonation alone is capable on breaking down the complex structures of organic matter (Equation 5). However, this reaction is selective (organic matter) and has a slow reaction rate [18].

$$O_3 + organic\ matter \rightarrow product \quad (5)$$

Based on the results presented in Figure 4, the effective pH condition can be concluded at pH 8 to 9. Additionally, the pH also ranges with the raw pH of SAL (pH 8.13) and gives an added advantage to this process as no pH adjustment is required during the treatment. Some of researcher also recorded the removal of COD up to 82% from dye wastewater under alkaline conditions [19]. More importantly, these results revealed that ozonation could potentially be used for remediating recalcitrant pollutants in SAL.

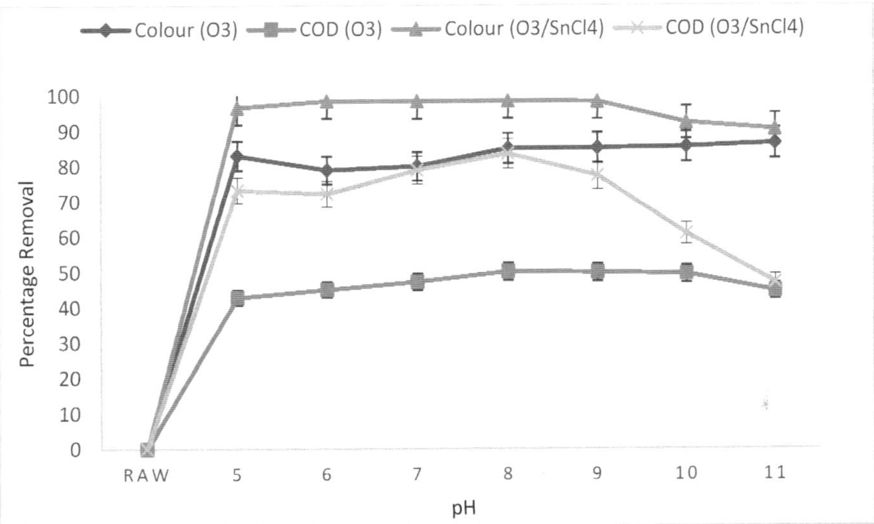

Figure 4. Effect of pH on colour and COD removal in SAL treatment during the O_3 and $O_3/SnCl_4$ process. (Experimental conditions: O_3 dosage 31 g/m^3, reaction time 60 minutes, gas flow rate 2.0 L/min, 500 mL of leachate sample)

3.2 Influence of dosage

The influence of O_3 and $SnCl_4$ dosage toward the removal is experimental and presented in Figure 5. The sample is constant at 500 mL while the pH and reaction time are fixed at raw pH 8.5 and 60 minutes, respectively, throughout the study. Briefly, it can be observed that the percentage of removals has consistently increased with the O_3 and $SnCl_4$ dosage for colour and COD. In the ozonation process (O_3) alone, the removal of colour and COD increase to 92% and 73%, accordingly, at a maximum O_3 dosage of 31 g/m^3 (Figure 5(a)). Meanwhile, a similar pattern is also depicted in Figure 5(b) for $O_3/SnCl_4$ treatment as the colour and COD removal graph increase with the dosage of $SnCl_4$ until 97% and 83%, respectively, before remaining constant afterwards. [20] described that the increment amount of catalyst would enhance the decomposition of ozone as well as pollutant removal. Moreover, the leachate characteristic, which is rich with dissolved organic matter (DOM) and has sufficient $OH\bullet^-$ generating moieties (i.e., Electron-rich-aromatic components), makes the decomposition of O_3 yield high with $OH\bullet$ concentration. This finding is also in line with previous scholars who recorded the removal of COD by the ozonation process as 37% to 50% [21] [22]. The differences might be due to the differences of operational condition applied (temperature, pollutant strength and gas flow rate). A comparison between ozonation alone and $O_3/SnCl_4$ shows that additional $SnCl_4$ has improved the performance of the ozonation, especially in the reduction of COD. This result is a line with [14] as reported that the performance of ozonation alone was poor compared to ozone/catalyst.

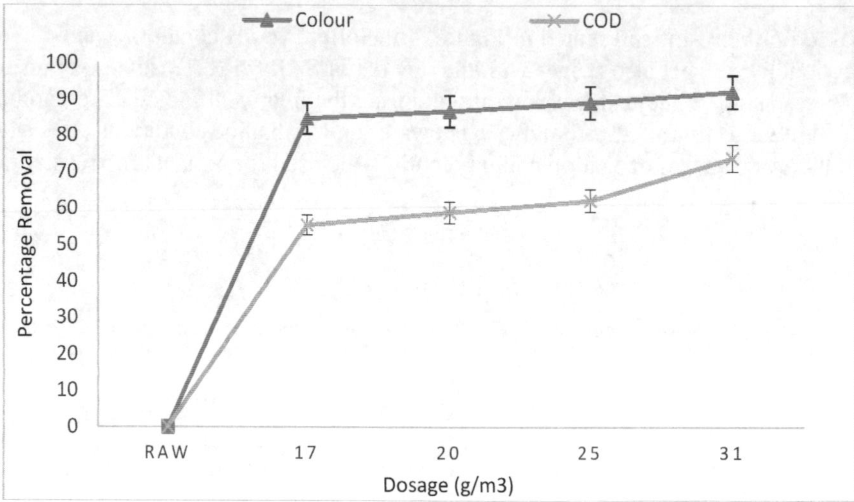

(a)

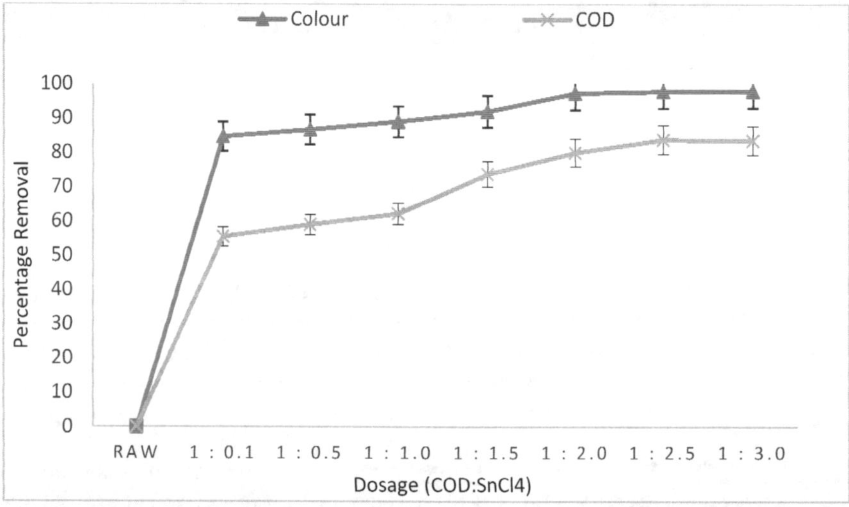

(b)

Figure 5. Effect of (a) O_3 dosage and (b) $SnCl_4$ dosage (ratio COD: $SnCl_4$) on colour and COD removal in SAL. (Experiment conditions: O_3 dosage 31 g/m^3 (for $O_3/SnCl_4$ test), raw pH, reaction time 60 min, gas flow rate 2 L/min, 500 mL of leachate sample).

3.3 Removal efficiency

Reaction time is another critical factor in physical and chemical reactions. Adequate time is required to maximise the reaction and determine the truth removal efficiency of the treatment. The reaction time range for O_3 and $O_3/SnCl_4$ was experimental and presented in Figure 6. According to Figure 6, removing colour and COD for ozonation alone starts to reach the optimal and equilibrium phase after the line hits at 30 min reaction time and gives maximum value as 85% and 50%, respectively. [23] explained that the equilibrium phase during the treatment is an indicator that organic matters are almost entirely susceptible to oxidation, and a few others can only be partially oxidised.

Meanwhile, $O_3/SnCl_4$ requires a longer time to achieve the optimal condition of 60 min compared to O_3. At this reaction time, the percentages of removal of colour and COD are recorded at 98% and 83%, respectively. Comparison between the removal of O_3 and $O_3/SnCl_4$ shows that leachate treatment using $O_3/SnCl_4$ is more efficient as it increases the removal up to 13% for colour and 40% for COD compared to O_3. This result also strengthens the hypothesis that additional $SnCl_4$ can potentially enhance the performance of the oxidation process. Thus, considering all the results obtained, the optimum time for O_3 and $O_3/SnCl_4$ to react in this study are 30 min and 60 min, respectively.

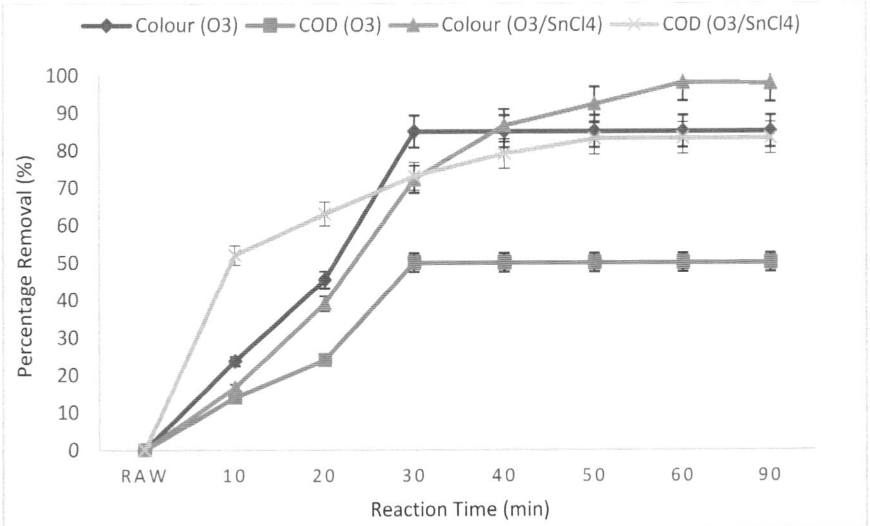

Figure 6. Effect of reaction time on colour and COD removal in SAL treatment during the O_3 and $O_3/SnCl_4$ process. (Experimental conditions: O_3 dosage 31 g/m^3, $SnCl_4$ dosage ratio ($SnCl_4$: COD) 1:1.25, pH 8, gas flow rate 2.0 L/min, 500 mL of leachate sample)

3.4 Biodegradability

The biodegradability ratio (BOD_5/COD) is another critical parameter of leachate that needs to be monitored. The biodegradation ratio is a crucial indicator to determine the period phase and describe the condition of leachate. In this section, the data recorded by a previous researcher and the current study (O_3 and $O_3/SnCl_4$) in improving BOD_5/COD are summarised in Table 3. Ozonation is proven as able to enhance the biodegradability of the sample since it can degrade organic matter into simpler structures. As depicts in Table 3, most of oxidation treatment has effectively improved the BOD_5/COD sample compared to other treatment types. For example, ozonation in this study has improved BOD_5/COD up to 0.23 from 0.08, while $O_3/SnCl_4$ enhances the biodegradability ratio from 0.03 to 0.28. [24] mentioned that almost 80% of the refractory organic compound was transferred into biodegradable materials after twenty minutes of microbubble ozonation. Other researchers have also recorded an improvement of BOD_5/COD after undergoing the ozonation process [25].

Table 3. Summary of biodegradability of SAL before and after treatment

Treatment method	BOD_5/COD		**References**
	Raw	After treatment	
O_3/H_2O_2	0.10	0.45	[26]
Fenton process	0.11	0.37	[27]
Pre-treatment using Air Stripping	0.12	0.10	[28]
Photo-Fenton	0.01	0.17	[29]
Ozonation	0.08	0.23	Current study
$O_3/SnCl_4$	0.03	0.28	Current study

4. Conclusion

According to the finding, the combination of ozone with $SnCl_4$ has the potential to be used as an effective leachate treatment and exhibits better removal performance of colour and COD up to 98% and 83%, respectively, at pH 8 and reaction time of 60 min. Moreover, $O_3/SnCl_4$ can improve the biodegradability ratio from 0.03 to 0.28. The optimal performance of this treatment is recorded at the alkaline condition, which indicates that the hydrolysis of $SnCl_4$ could enhance the production of OH• and accelerate the degradation of pollutants using an indirect pathway. However, further investigation is recommended to understand the kinetic reaction of $O_3/SnCl_4$ and further optimise the $SnCl_4$ dosages (using Respond Surface Methodology software (RSM)) from the current study as it is still high. The RSM is suggested as this software capable to find the ideal process in optimising the experimental conditions as well as pollutant removal precisely, through mathematical and statistical techniques (ANOVA).

References

[1] Krzemińska D, Neczaj E and Borowski G 2015 Advanced oxidation processes for food industrial wastewater decontamination. *Journal of Ecological Engineering*. **16** (2).

[2] Amr S S, Zakaria S N F and Aziz H A 2017 Performance of combined ozone and zirconium tetrachloride in stabilized landfill leachate treatment. Journal of Material cycles and waste management, Oct;**19** (4):1384-90.

[3] Amr S S A, Aziz H A, Adlan M N and Bashir M J 2013 Pretreatment of stabilized leachate using ozone/persulfate oxidation process. *Chemical Engineering J*. **221**, 492- 499

[4] Zhu S N, Hui K N, Hong X and Hui K S 2014 Catalytic ozonation of basic yellow 87 with a reusable catalyst chip. *Chemical engineering J*, 242, 180-186.

[5] Amr S S A, Aziz H A, Hossain M S and Bashir M J 2017 Simultaneous removal of COD and color from municipal landfill leachate using Ozone/Zinc Sulphate oxidation process. *Global NEST J*., **19** (3), 498-504.

[6] Ramdhani M Y, Sururi M R and Ainun S 2018 Leachate Treatment from Sarimukti Landfill Using Ozone with Sludge from Water Treatment Plant as a Catalyst. In MATEC Web of Conferences (Vol. 147, p. 04006). EDP Sciences.

[7] Batakliev T, Georgiev V, Anachkov M and Rakovsky S 2014 Ozone decomposition. *Interdisciplinary toxicology*, 7(2), 47-59.

[8] Chen D, Wang D, Wu W and Xiao L 2015 SnCl4· 5H2O: A Highly Efficient Catalyst for

Hydration of Alkyne. *Applied Sciences*, **5** (2), 114-121.
[9] Zainal, S. F. F. S., & Aziz, H. A. (2017, October). Potential of tin (IV) chloride for treatment in Alor Pongsu as stabilized landfill leachate. In AIP Conference Proceedings (Vol. 1892, No. 1, p. 040003). AIP Publishing.
[10] American Public Health Association (APHA) 2012 Standard methods for the examination of water and wastewater. Washington, DC, USA.
[11] Zakaria S N, Aziz H A, Alazaiza M Y 2022 Effectiveness of ozonation with zirconium and tin tetrachloride for stabilized anaerobic landfill leachate treatment. Water Environment Research. 2022 Jan;94(1):e1672..
[12] Zakaria, S. N. F. (2020). Treatment of Stabilised Anaerobic Landfill Leachate by Ozonation Process with Zirconium and Tin Tetrachloride, PhD thesis, Universiti Sains Malaysia.
[13] Sarif S M, Alias S S, Ridwan F M, Salim K K, Abidin C Z A and Ali U M 2018 Ozonation of return activated sludge for disintegration and solubilisation with synthesised titanium oxide as catalyst. In E3S Web of Conferences. 34, 02009. EDP Sciences. https://doi.org/10.1051/e3sconf/20183402009.
[14] Amr S S A, Zakaria S N F and Aziz H A 2016 Performance of combined ozone and zirconium tetrachloride in stabilised landfill leachate treatment. *J Mater Cycles Waste Manage*. **19** (4), 1384-1390. https://doi.org/10.1007/s10163-016-0524-x.
[15] Li W, Zhou Q and Hua T 2010 Removal of organic matter from landfill leachate by advanced oxidation processes: a review. *Int J of Chemical Engineering*, 2010.
[16] Wang J L and Xu L J 2012 Advanced oxidation processes for wastewater treatment: formation of hydroxyl radical and application. Critical reviews in environmental science and technology, 42(3), 251-325.
[17] Akatova I N, Nikulin S S, Korystin S I, Kondrat'eva N A 2004 Effect of the type of coagulant on coagulation properties of rubbers, compounds and vulcanisates. International Polymer Science and Technology. Aug;31(8):1-4.
[18] Guo Y, Yang L, Cheng X and Wang X 2012 The application and reaction mechanism of catalytic ozonation in water treatment. *J Environ Anal Toxicol*, **2** (7), 2161-0525.
[19] Multani M Y, Shah M J and Student P G 2014 Removal of colour and COD from reactive green–19 dyeing wastewater using ozone. Int. *J. Eng. Sci. Res. Technol*, **3** (2), 699-704.
[20] Wang H, Zhan J, Yao W, Wang B, Deng S, Huang J and Wang Y 2018 Comparison of pharmaceutical abatement in various water matrices by conventional ozonation, peroxone (O3/H2O2), and an electro-peroxone process. *Water research*, 130, 127-138.
[21] Frontistis Z, Xekoukoulotakis N P, Diamadopoulos E and Mantzavinos D 2008 Ozonation of landfill leachates: Treatment optimization by factorial design. *J.Advanced Oxidation Technologies*, 11(2), 370-376.
[22] Tizaoui C, Bouselmi L, Mansouri L and Ghrabi A 2007 Landfill leachate treatment with ozone and ozone/hydrogen peroxide systems. *J. hazardous materials*, 140 (1-2), 316-324.
[23] Muhammad A, Shafeeq A, Butt M A, Rizvi Z H, Chughtai M A and Rehman S 2008 Decolorization and removal of cod and bodfrom raw and biotreated textile dye bath effluent through advanced oxidation processes (AOP). Brazilian *J of Chemical Engineering*, 25(3), 453-459.
[24] Liu S, Wang Q, Zhai X, Huang Q and Huang P 2010 Improved Pretreatment (Coagulation-Floatation and Ozonation) of Younger Landfill Leachate by Microbubbles. *Water Environment Research*, 82(7), 657-665.
[25] Ulucan-Altuntas K and Ilhan F 2018 Enhancing biodegradability of textile wastewater by ozonation processes: Optimization with response surface methodology. Ozone: *Science & Engineering*, 40(6), 465-472.
[26] Leszczyński J and Maria J W 2018 The removal of organic compounds from landfill leachate using ozone-based advanced oxidation processes. *In E3S Web of Conferences* (Vol. 45, p.

00046). EDP Sciences.
[27] Bernardo-Bricker A R, Singh S K, Trovó A G, Tang W Z and Tachiev G 2014 Biodegradability enhancement of mature landfill leachate using Fenton process under different COD loading factors. *Environmental Processes*, 1 (3), 207-219.
[28] Poveda M, Yuan Q and Oleszkiewicz J 2016 The effectiveness of pretreatment methods on COD and ammonia removal from landfill leachate. *International Journal of Environmental Science and Development*, 7 (4), 257.
[29] Zazouli M A, Yousefi Z, Eslami A and Ardebilian M B 2012 Municipal solid waste landfill leachate treatment by fenton, photo-fenton and fenton-like processes: Effect of some variables. *Iranian J. environmental health science & engineering*, 9 (1), 3.

AUTHOR INDEX